A SHORT GUIDE TO CLIMATE CHANGE RISK

SHORT GUIDES TO RISK SERIES

Risk is a far more complex and demanding issue than it was ten years ago. Risk managers may have expertise in the general aspects of risk management and in the specifics that relate directly to their business, but they are much less likely to understand other more specialist risks. Equally, Company Directors may find themselves falling down in their duty to manage risk because they don't have enough knowledge to be able to talk to their risk team in a sensible way.

The short guides to risk are not going to make either of these groups experts in the subject but will give them plenty to get started and in a format and an extent (circa 150 pages) that is readily digested.

Titles in the series will include:

- Climate Change Risk
- Compliance Risk
- Employee Risk
- Environmental Risk
- Fraud Risk
- Information Risk
- Intellectual Property Risk
- Kidnap and Ransom Risk
- Operational Risk
- Purchasing Risk
- Reputation Risk
- Strategic Risk
- Supply Chain Risk
- Tax Risk
- Terrorism Risk

A Short Guide to Climate Change Risk

Nigel Arnell

Routledge
Taylor & Francis Group
LONDON AND NEW YORK

First published 2015 by Gower Publishing

Published 2016 by Routledge
2 Park Square, Milton Park, Abingdon, Oxon OX14 4RN
711 Third Avenue, New York, NY 10017, USA

Routledge is an imprint of the Taylor & Francis Group, an informa business

British Library Cataloguing in Publication Data
A catalogue record for this book is available from the British Library

The Library of Congress has cataloged the printed edition as follows:
Arnell, Nigel, author.
A short guide to climate change risk / by Nigel Arnell.
pages cm. -- (Short guides to business risk)
Includes bibliographical references and index.
ISBN 978-1-4094-5352-9 (paperback : alk. paper) -- ISBN 978-1-4094-5353-6 (ebook) -- ISBN 978-1-4724-0803-7 (epub)
1. Climatic change--Risk assessment. 2. Climatic change--Environmental aspects. 3. Climatic change--Economic aspects. 4. Climatic change--Effect of human beings on I. Title.

QC903.A76 2015
363.738'742--dc23

2014029432

ISBN 13: 978-1-4094-5352-9 (pbk)

Contents

List of Figures

List of Tables

List of Acronyms

BACLIAT	Business Area Climate Impacts Assessment Tool
CCC	(UK) Committee on Climate Change
CCRA	(UK) Climate Change Risk Assessment
CCS	Carbon Capture and Storage
CDM	Clean Development Mechanism
CO_2	Carbon Dioxide
COP	Conference of the Parties (to the UNFCCC)
ENSO	El Niño/Southern Oscillation
ETS	European Union Emissions Trading Scheme
GCM	Global Climate Model
GEF	Global Environmental Facility
IPCC	Intergovernmental Panel on Climate Change
JI	Joint Implementation
LCLIP	Local Climate Impacts Profile
LULUCF	Land Use, Land Use Change and Forestry
MCA	Multi-Criteria Analysis
NAMA	Nationally Appropriate Mitigation Action
NAO	North Atlantic Oscillation
NAPA	National Adaptation Programme of Action
RCP	Representative Concentration Pathway
REDD	Reducing Emissions from Deforestation and forest Degradation

ROA	Real Options Analysis
SBI	Subsidiary Body on Implementation (to the UNFCCC)
SBSTA	Subsidiary Body on Scientific and Technical Advice (to the UNFCCC)
SPM	Summary for Policymakers (IPCC)
SRES	Special Report on Emissions Scenarios
UKCIP	United Kingdom Climate Impacts Programme
UKCP09	United Kingdom Climate Projections 2009
UNDP	United Nations Development Programme
UNEP	United Nations Environment Programme
UNFCCC	United Nations Framework Convention on Climate Change
WMO	World Meteorological Organisation

Acknowledgements

This book is informed by more than two decades of research into climate change and its consequences, during which I have worked not only with students and academic colleagues from across the world, but also with government departments, public sector agencies and many private sector companies. Their comments and questions have stimulated new ideas, and forced me to think carefully about how to assess the risks posed by climate change and communicate information on climate change to diverse audiences. I would particularly like to thank my colleagues at the Walker Institute for Climate Systems Research at the University of Reading, especially Dr Maria Noguer, Kathy Maskell and Dr Ben Lloyd-Hughes.

For my wife, Dr Hilary Arnell, who has supported and encouraged me not only whilst I have been writing this book, but during all my endeavours over the years to understand and communicate climate change risks: I am truly grateful for her advice and encouragement.

1 Climate Change as a Business Risk

INTRODUCTION: WHAT IS CLIMATE CHANGE?

Over the last few years the phrase 'climate change' has hardly been out of the media. A succession of extreme weather events around the world – including Tropical Storm Sandy causing widespread damage in New York in 2012, Typhoon Haiyan hitting the Philippines in 2013 and several bouts of extreme wet and hot weather in Europe – have illustrated how vulnerable both developed and developing economies are to extreme weather. Politicians have devoted time and effort to try to negotiate an international climate change treaty. Campaign groups urge them to do more to avert major catastrophe, but others suggest that the threats posed by climate change are overstated. Governments have tried several – controversial – ways of encouraging us to use energy more efficiently and to increase the use of renewable energy sources. Increasingly, climate change is becoming relevant for many organisations in both the private and public sectors. It affects both how they use energy and its price, it affects the risks to which an organisation is exposed, and can also open up new opportunities.

The most recent reports of the United Nations Intergovernmental Panel on Climate Change (IPCC) were released in late 2013[1] and early 2014.[2, 3] Their headline conclusions were:

- Warming of the climate system is unequivocal, and each of the last three decades has been successively warmer than any decade since 1850.
- Carbon dioxide concentrations in the atmosphere have increased by 40% since pre-industrial times, and about half of cumulative human CO_2 emissions since 1750 have occurred in the last 40 years.
- It is extremely likely that this human influence has been the dominant cause of the observed warming since the mid 20th century.
- Continued emissions of greenhouse gases will cause further warming and changes in the climate system. Without increased efforts to reduce emissions, global mean surface temperature will be 2.5-7.8°C above pre-industrial levels by 2100.
- Climate change increases risks from extreme events (such as storms, floods, heatwaves and droughts), but these risks vary from place to place.

---

1 IPCC (2013), Summary for Policymakers. In: *Climate Change 2013: The Physical Science Basis. Contribution of Working Group I to the Fifth Assessment Report of the Intergovernmental Panel on Climate Change*. Stocker, T., et al., eds. Cambridge University Press.
2 IPCC (2014), Summary for Policymakers. In: *Climate Change 2014: Impacts, Adaptation and Vulnerability. Contribution of Working Group II to the Fifth Assessment Report of the Intergovernmental Panel on Climate Change*. Field, C., et al., eds. Cambridge University Press.
3 IPCC (2014), Summary for Policymakers. In: *Climate Change 2014: Mitigation of Climate Change. Contribution of Working Group III to the Fifth Assessment Report of the Intergovernmental Panel on Climate Change*. Edenhofer, O., et al., eds. Cambridge University Press.

- Limiting climate change will require substantial and sustained reductions of greenhouse gas emissions. This will reduce the overall risks of climate change impacts.

- Responding to climate-related risks involves decision-making in a changing world, with continuing uncertainty about the severity and timing of climate change impacts and with limits to the effectiveness of adaptation.

The aim of this book is to explain the science behind climate change, to discuss the potential impacts, and to describe ways of addressing the risks and opportunities posed by climate change. It seeks to provide answers to questions such as:

- Can we really see any effects of climate change?

- Can we be confident that the climate will change in the future?

- What is the difference between weather variability and climate change?

- How might climate change in the future, and how will it affect my organisation?

- How can we adapt to climate change risks and exploit opportunities?

CLIMATE CHANGE AND BUSINESS RISK

The World Economic Forum's Global Risk 2014 report[4] surveyed over 700 global business leaders and decision-makers

4 World Economic Forum (2014), *Global Risks 2014*. World Economic Forum, Geneva.

on 31 global risks. Two of the top ten risks directly relate to climate change: the failure of climate change policy came in at number 5, and number 6 was a greater incidence of extreme weather events. Two of the other risks – water crises (at number 3) and food crises (number 8) – are also affected by climate change. In the UK, a survey of the FTSE 350 companies in 2013[5] showed that 86% of companies had identified potential risks from climate change; 84% had identified risks from regulations, 74% had identified physical risks, and 70% had identified other risks such as taxes and changes in behaviour. The five most commonly reported risks related to tax, regulation and consumer behaviour. 82% of the companies identified potential opportunities from climate change – again these mostly related to regulations, taxes and behaviours and only 59% identified potential opportunities from changes in climate itself. An earlier UK report[6] found that whilst 73% of large companies were 'thinking' about climate change, only 56% of small companies were doing so, and less than a quarter were actually doing something. 34% of the companies surveyed were 'very concerned' about the effects of climate change on the UK.

There is therefore some awareness of climate change amongst the business community, but this varies considerably between organisations. The perceptions of the nature of risks also vary, reflecting the many ways in which climate change may affect an organisation. There are five broad areas of risk from climate change:

- Operational: climate change may affect the ability or costs of doing business;

5 CDP (2013), *Are UK businesses prepared for international impacts of climate change?* FTSE 350 Climate Change Report 2013. CDP, London.
6 IPSOS-MORI (2010), *Climate Change Adaptation: A Survey of Private, Public and Third Sector Organisations*. Report to Defra.

- Supply chains and procurement: climate change may impact upon an organisation's supply chain;
- Market: changes in climate may alter the demand for an organisation's product or service;
- Regulatory: changes in climate may lead to new regulations which constrain – or facilitate – an organisation's activities;
- Reputation: the way an organisation is perceived to be responding to climate change may affect its reputation.

An organisation should therefore consider adding climate change to its risk register.

ADAPT TO IMPACTS AND OPPORTUNITIES, OR REDUCE EMISSIONS?

There are two ways to address the risks posed by climate change. The first is to seek to reduce the amount of climate change that will occur. This could be achieved by slowing or reversing the increase in concentration of greenhouse gases in the atmosphere, which is known as 'mitigation' in the language of climate change.[7] These reductions in concentration could be achieved in several ways. The most obvious is through using less fossil fuel by using less energy, increasing energy efficiency or by the substitution of other sources of energy. Greenhouse gases can also in principle be extracted from waste and stored so that they do not accumulate in the atmosphere; carbon capture and storage (CCS) does this

7 It is important to note that this is a much narrower interpretation of a term that is more generally used to apply to measures to reduce risk more generally.

by extracting CO_2 from the effluent from power stations and burying it deep underground. It is also possible to reduce the increase in emissions by slowing the removal of forest cover (deforestation releases carbon into the atmosphere), and to increase the withdrawal of carbon from the atmosphere by planting trees. More controversially, a range of fanciful 'geoengineering' approaches have been proposed to extract CO_2 from the atmosphere by fertilising the oceans to absorb more CO_2, or to offset the effects of increasing greenhouse gases by reflecting solar radiation back into space. These face a multitude of practical, conceptual and political challenges, and are highly unlikely to be implemented in practice.

The second approach is to seek to adjust to changing climate risks and take advantage of new opportunities: this is known as 'adaptation'. Some adaptation will take place autonomously as organisations react to changing stimuli, but much will need to be planned. There are many potential ways of adapting to climate change, as we shall see later. Organisations can adapt specifically to climate change, or can incorporate climate change as just another driver into normal business planning.

In practice, an organisation's response to climate change is likely to contain elements of both mitigation and adaptation, and there are of course linkages between measures to mitigate climate change and measures to adapt to changing risks. Mitigation and adaptation objectives and measures may in some cases conflict, and in other cases complement each other. For example, a strategy to adapt to warmer temperatures by increasing air conditioning would conflict with a strategy to reduce energy use. On the other hand, it may be possible to design a new building, for example, that both used less energy and was less prone to overheating.

National and international climate change policies have tended to concentrate on mitigation, although they are increasingly concerned with supporting adaptation.

CLIMATE CHANGE AND SUSTAINABILITY

Back in 1987 the World Commission on Environment and Development (also known as the Brundtland Commission) published 'Our Common Future',[8] which introduced the concept of sustainable development:

> *Sustainable development is development that meets the needs of the present without compromising the ability of future generations to meet their own needs.*

Since then, there have been many attempts to implement sustainable development amongst businesses in practice, and many guides have been written[9] emphasising the commercial benefits of sustainability: these include lower costs and an enhanced 'licence to operate'. The World Business Council on Sustainable Development published a vision for 2050 for a world 'on the way to sustainability'.[10]

The devil is in the detail – and the concept of sustainability is intrinsically ambiguous – but most attempts at implementing sustainable development in an organisation focus on balancing the four capitals: financial, human, social and natural. 'Natural capital' represents the environmental

8 World Commission on Environment and Development (1987), *Our Common Future*. Oxford University Press.

9 For example Blowfield, M. (2012), *Business and Sustainability*. Oxford University Press; Weybrecht, G. (2013), *The Sustainable MBA: A Business Guide to Sustainability*. Second Edition, Wiley.

10 World Business Council for Sustainable Development (2010), *Vision 2050: The New Agenda for Business*. WBCSD, Geneva.

resources that are used in the production of goods and services, and the environmental resources and qualities that are depleted by their production and use. Overexploitation and pollution of the natural environment deplete natural capital. The key difficulties in implementing sustainable development practices lie in the characterisation and balancing of the four capitals.

Acting 'sustainably', however, does not in itself mean that an organisation is addressing climate change. Climate change adds an extra dimension. It influences the availability of natural capital (including through exposure to hazard), and the priorities for sustainable development affect the emissions of greenhouse gases and vulnerability to climate change.

ABOUT THIS BOOK

This book presents an overview of the science of climate change and the risks posed by a changing climate, and introduces methods for assessing and responding to these risks. The book is aimed at those people in organisations who need to know about climate change as a strategic and operational risk. 'Organisation' is defined loosely to include public and third sector bodies as well as private companies; their needs and expectations will of course be very diverse, so the book focuses on general principles rather than specific measures or procedures. The focus is on understanding and managing climate change risks, and not on the specifics of how to reduce greenhouse gas emissions.

Chapter 2 introduces the science of climate change in more detail, presenting the physical principles behind our understanding of climate change and explaining how we can make projections of climate into the future. It introduces the major sources of uncertainty in our projections, and this uncertainty is crucial in understanding how to respond to climate change. Chapter 3 explores the potential impacts of climate change on the economy, society and the environment, and also on organisations. It explores adaptation to climate change, and the barriers that it faces.

Chapter 4 provides an overview of international and national climate policy. This is necessary because the future impacts of climate change depend on these policies, and the way organisations can respond to climate change will be facilitated or constrained by policy. The chapter also introduces some initiatives from within the business community and its regulators that affect response to climate change.

Chapter 5 introduces risk management frameworks, and describes how the risks posed by climate change to an organisation can be assessed; it discusses how climate scenarios are constructed and can be applied. Chapter 6 discusses how organisations can decide how to cope with climate change risks and opportunities, given the uncertainties in projected future climate change. Chapter 7 concludes by drawing some key messages from the preceding chapters.

Each chapter ends with a summary of the key points. Climate change science and policy is riddled with acronyms, so a list of acronyms is provided.

CONCLUSION: SOURCES OF INFORMATION

Although the basic physics behind climate change have been known since the middle of the 19th century, the science of climate change is developing rapidly. Most of this development is presented in scientific papers in the academic literature. Fortunately, every few years the Intergovernmental Panel on Climate Change (IPCC) publishes an authoritative, in-depth assessment of the state of the knowledge of climate change and how we can respond to it.

The IPCC[11] was created in 1988 as an inter-governmental body overseen by the World Meteorological Organisation and the United Nations Environment Programme, and charged with providing periodic assessments to the United Nations and member states; 195 countries are currently members of the IPCC. The IPCC operates by convening Working Groups, which cover the science of climate change (Working Group I), impacts, adaptation and vulnerability (Working Group II) and the mitigation of climate change (Working Group III). It also conducts occasional special reports, most recently on renewable energy[12] and on climate change and extreme events.[13]

Each Working Group consists of groups of scientists and experts who assess the available relevant scientific information, in a highly iterative process involving review by other experts and governments. Each Working Group produces a high level 'Summary for Policymakers' (SPM), which is agreed line-by-line by government representatives in often fraught week-long negotiating sessions. Underpinning each SPM is a large volume containing the full assessment, and these volumes provide an excellent authoritative source of information on climate change.

11 www.ipcc.ch.

12 IPCC (2011), *Renewable Energy Sources and Climate Change Mitigation*. Edenhofer, O., et al. (eds), Cambridge University Press.

13 IPCC (2012), *Managing the Risks of Extreme Events and Disasters to Advance Climate Change Adaptation*. Field, C., et al., eds. Cambridge University Press.

The first IPCC report was completed in 1990, with the second, third and fourth produced in 1995, 2001 and 2007 respectively. The most recent Fifth Assessment Report was completed between September 2013 and April 2014. Since the third report, a synthesis report has also been produced drawing together the assessments from all three working groups.

An increasing number of public agencies, business organisations and individual companies have produced advice and guidance on coping with the risks of climate change. Many reports are referenced in this book. Much useful information and guidance is also available online, and can easily be found through search engines (web addresses rapidly become out of date). This book provides a framework for searching for, and interpreting, this increasing mass of information.

2 The Science of Climate Change

INTRODUCTION

'climate is what you expect, weather is what you get'

We are all very familiar with the notion that climate varies from place to place. The south of England is usually warmer than northern Scotland, and considerably cooler than the south of Spain. The earth's climate system is driven by energy from the sun, which warms the surface and evaporates water. Differences in climate arise not only because the amount of energy received varies from place to place, but also because the earth's atmosphere and oceans transport energy and water from one place to another. The weather we experience on any one day is therefore determined by the interaction between time of year and the movement of these two fluids. The weather across the British Isles is typically characterised by a sequence of low pressure depressions moving eastwards from the Atlantic Ocean, bringing rain and cloud, and the warm waters of the Atlantic flowing from the south west keeping temperatures mild. The amount of rain that falls varies depending on how intense these depressions are, and how far north they track. Sometimes the weather of the British Isles is dominated by high pressure over Europe and Asia. In winter this tends to produce cold, dry weather, whilst in summer it usually produces hot, dry weather.

The distinction between weather and climate is illustrated in Figure 2.1. The left side shows the average temperature and rainfall through the year in the town of Reading, in Berkshire in southern England, and if we just experienced 'climate' this is what we would get. The right side shows actual daily temperatures and monthly rainfall in 2011; there is clearly a lot of variability in weather around the average climate. Weather also varies from year to year. This variability is a characteristic of a climate regime at a place, and in some parts of the world there is greater variability from year to year than in others. Rainfall in dry regions tends to vary greatly from year to year.

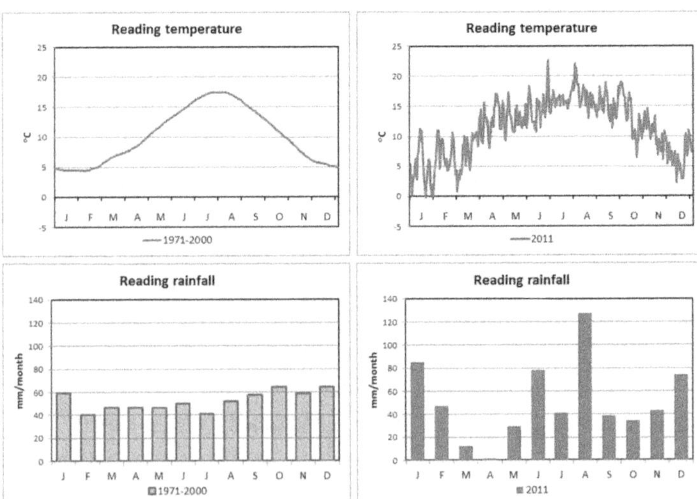

Figure 2.1 **Weather and climate: temperature at Reading, UK. The data were collected from the Atmospheric Observatory at the University of Reading**

Climate change can be defined rather formally as a change in the statistical properties of weather and climate at a place, in terms of average, variability or both. These properties obviously have to be calculated over a particular time period, and by convention climatologists use a 30-year period to define climate. Whilst the number of interesting weather events (such as high temperature extremes) may vary substantially from one five-year period (say) to another, it is usually extremely difficult to distinguish the statistical properties of weather between such short periods. Climate change can therefore be seen as representing a long-term shift in climate at a place.

This definition of climate change, of course, begs the question: 'change relative to what?' This 'baseline' varies between applications, and between different areas of climate change policy and response. The IPCC's Working Group I report[1] uses both 1961-1990 and 1986-2005 as baseline periods, and many – but not all – studies of the potential consequences of climate change (Chapter 3) use 1961-1990 to define baseline climate. However, international climate policy (Chapter 4) uses a 'pre-industrial' baseline to set temperature targets, and this is variously defined as mid-18th century, 1850 or the period 1850-1900. There are good reasons why baselines vary – largely related to the availability of information – but it is important to be aware that the selection of the baseline influences the interpretation of the amount of climate change that has occurred and will occur in the future.

The rest of this chapter outlines why and how climate may change. For a number of reasons – examined in detail below

1 IPCC (2013), Summary for Policymakers. In: *Climate Change 2013: The Physical Science Basis. Contribution of Working Group I to the Fifth Assessment Report of the Intergovernmental Panel on Climate Change.* Stocker, T., et al., eds. Cambridge University Press.

– we do not know exactly how climate will change in the future, and this is one of the key challenges in responding to the risks and opportunities posed by climate change.

IS CLIMATE CHANGING, AND WHY?

Since the 1970s the average temperature of the surface of the earth has increased by around 0.5°C, and temperatures in the first decade of the 21st century have been around 0.8°C higher than in the middle of the 19th century[2] (Figure 2.2). The increases in temperature have been higher over the land than over the sea, and across most of the land temperatures are now on average more than 1°C higher than in the 1970s. There are several possible explanations for this.

First, the apparent trend could be artificial – it could be an artefact of the way temperature records have been collected and analysed. It is well known, for example, that cities are warmer than surrounding areas, and that as cities grow this 'heat island' effect increases. However, the data that go into the global calculations are corrected for this bias, and different ways of aggregating the observations – which of course are not equally spread across the world – into global averages produce broadly similar trends. The change in global average surface temperature shown in Figure 2.2 is therefore real.

The earth's climate is driven by the sun, so variations in solar output of energy could therefore perhaps be expected to

2 IPCC (2013), Summary for Policymakers. In: *Climate Change 2013: The Physical Science Basis. Contribution of Working Group I to the Fifth Assessment Report of the Intergovernmental Panel on Climate Change*. Stocker, T., et al., eds. Cambridge University Press.

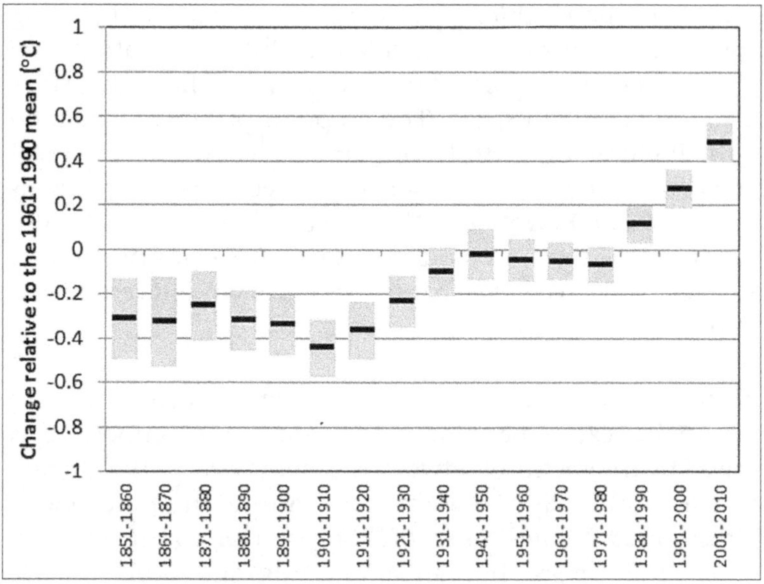

Figure 2.2 Change in global mean surface temperature, 1851-2010. The data come from the HadCRUT4 data set, and the grey bars represent the uncertainty in estimated temperature change

lead to variations in climate on earth. The energy output by the sun varies over the 11-year sunspot cycle, but although there have been many attempts to find matching variations in weather and climate, current evidence suggests a relatively small effect at the global scale (although perhaps with larger effects in some regions and some seasons). Over the much longer term, the radiation received by the earth varies with the shape of the earth's orbit and the orientation of its axis. These variations, with cycles of around 20,000, 40,000 and 100,000 years, are associated with ice ages, and are far too slow to explain the rise in global mean temperature over the last few decades.

Some volcanic eruptions can cool the climate by injecting large volumes of sulphur dioxide high into the stratosphere; this condenses to form aerosols[3] which reflect the sun's radiation back into space. The eruption of Mount Pinatubo in the Philippines in 1991, for example, led to a fall in global temperature of up to 0.5°C for two or three years, and previous eruptions from Krakatoa and Tambora in Indonesia in the 19th century and Laki in Iceland in the late 18th century also had large regional effects. The consequences of volcanic eruptions, however, are short-lived.

Internal variability within the climate system can also potentially cause periodic warm and cool periods. It is caused by interactions between different parts of the climate system, most usually between the ocean and the atmosphere. The tropical Pacific Ocean off the coast of South America, for example, goes through a warming and cooling cycle approximately three times a decade, and this is linked with variability in the atmosphere (known as the El Niño/Southern Oscillation, or ENSO). The warm phase – El Niño – is associated with higher than average rainfall in parts of the western United States and South America, lower rainfall in eastern Australia and southern Africa, and may affect weather even further afield. The temperature of the North Atlantic Ocean also varies over time (known as the Atlantic Multidecadal Oscillation), with fluctuations between sequences of warm and cool years. The North Atlantic Oscillation (NAO) is a periodic fluctuation in pressure gradients across the Atlantic Ocean, which affects the location and strength of weather patterns crossing into Europe. Both affect temperature, rainfall and wind over Europe and parts of North America. However, whilst such internal variability can explain variations in regional weather from year to year and, in some

3 Small particles.

places, decade to decade, it cannot explain the long-term increase in average temperature shown in Figure 2.2.

This long-term increase can only be explained by invoking another mechanism – changes in the composition of the atmosphere due to human activities. This is variously known as 'the enhanced greenhouse effect' or 'anthropogenic global warming', and whilst it has only recently become a major concern, our understanding of the fundamental physics dates back to the early 19th century.

THE PHYSICS OF CLIMATE CHANGE

The earth receives radiation from the sun, and radiates an equivalent amount of energy back out to space. Given the amount of energy received from the sun, and accounting for radiation reflected back into space, the average temperature at the surface of the earth should be approximately -18°C. However, the actual average temperature is around 15°C, and this is because the earth has an atmosphere. The sun emits most of its energy as shortwave radiation, but the earth – being cooler than the sun – emits most of its energy at longer wavelengths. The earth's atmosphere is effectively transparent to the incoming shortwave radiation, but some of the longwave radiation emitted from the surface is absorbed by gases in the atmosphere and re-radiated in all directions, including downward. The surface of the earth therefore warms. This is known as the 'greenhouse effect',[4] and was first suggested by the French mathematician Joseph Fourier back in 1824. In the late 1850s John Tyndall demonstrated that water vapour, carbon dioxide and some other gases absorbed radiant

4 Although this is a misnomer, as the main reason why temperatures are higher in a real greenhouse is because the movement of air is reduced.

heat, and at the end of the 19th century the Swedish chemist Svante Arrhenius suggested that changes in the concentration of carbon dioxide in the atmosphere could lead to changes in global temperature by increasing the amount of energy at the surface of the earth.

Carbon dioxide (CO_2) is released into the atmosphere by the decomposition of plant material, and absorbed by plants and the oceans. Deforestation leads to an increase in CO_2 concentration, partly because of the decomposition and partly because the cleared forests no longer absorb CO_2. The combustion of fossil fuels (coal, gas and oil) releases carbon from long-term storage. Figure 2.3 shows the increase in emissions of CO_2 from deforestation and fossil fuel burning between 1970 and 2010;[5] the total has nearly doubled over 40 years, and the increase from 2000 to 2010 was higher than in the previous three decades.

CO_2 concentrations in the atmosphere have increased from around 315 ppmv[6] in the late 1950s (when systematic measurements first started) to approximately 391 ppmv in 2011[7] (Figure 2.4). Most of this increase is due to the combustion of fossil fuels, and it is estimated that the CO_2 concentration was around 280 ppmv before industrialisation (in other words, it has increased by around 40% since the mid-19th century). Human activity has also increased the concentration in the atmosphere of other greenhouse gases,

5 IPCC (2014), Summary for Policymakers. In: *Climate Change 2014: Mitigation of Climate Change. Contribution of Working Group III to the Fifth Assessment Report of the Intergovernmental Panel on Climate Change*. Edenhofer, O., et al., eds. Cambridge University Press.

6 Parts per million by volume.

7 IPCC (2013), Summary for Policymakers. In: *Climate Change 2013: The Physical Science Basis. Contribution of Working Group I to the Fifth Assessment Report of the Intergovernmental Panel on Climate Change*. Stocker, T., et al., eds. Cambridge University Press.

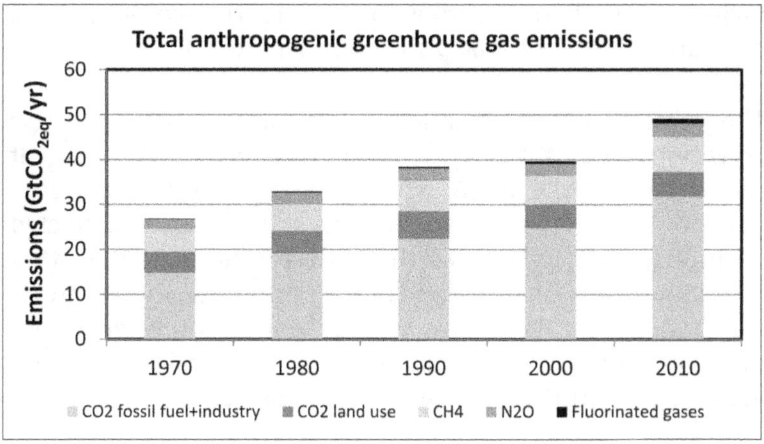

Figure 2.3 **Total anthropogenic emissions of greenhouse gases: 1970 to 2010. From the IPCC AR5 Working Group III Summary for Policymakers**

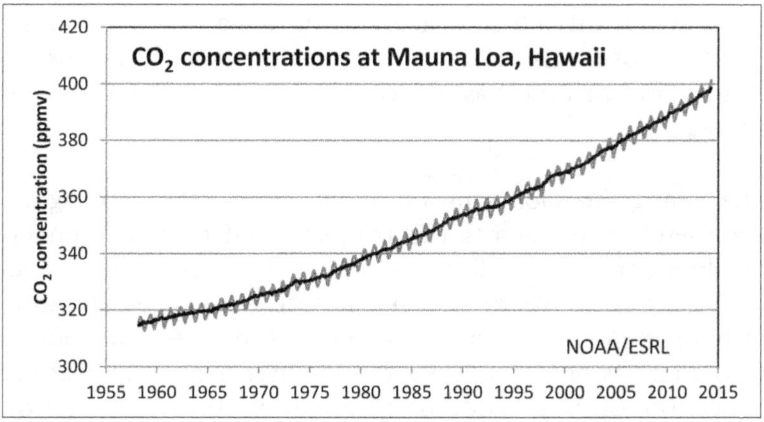

Figure 2.4 **CO_2 concentrations in the atmosphere at Mauna Loa in Hawaii. Data from NOAA/ESRL**

including methane, nitrous oxide, fluorinated gases and halocarbons: these are also shown in Figure 2.3.[8]

The increasing concentration in the atmosphere of these greenhouse gases leads to an increase in temperature at the surface of the earth. The most important greenhouse gas, however, is water vapour. This is not directly affected by human activity but, as the concentrations of the other greenhouse gases increase and temperatures rise, more water is evaporated from the oceans and the amount of water vapour in the atmosphere rises. This acts as a positive feedback, and increases still further the rise in temperature.

The burning of fossil fuels also leads to the emission of aerosols into the atmosphere. Aerosols affect the absorption and scattering of radiation, and although different aerosols have different effects, the net effect of increasing aerosols since the mid-20th century has been to *reduce* energy at the surface. This has therefore offset to a small extent the enhanced greenhouse effect. Aerosols also affect the formation and brightness of clouds, and this too has reduced the energy available to the surface.

The change in global mean surface temperature in Figure 2.2 therefore represents the combined effect of a number of different causes. The dominant driver has been the increasing concentration of greenhouse gases. The flattening of temperatures between 1940 and 1970 has been attributed to the effects of increasing aerosols offsetting the increasing concentration of greenhouse gases (before pollution control

8 The effect of these other greenhouse gases on the radiative balance are expressed in terms of the effect of the same amount of CO_2, and they are combined to produce an estimate of the 'equivalent CO_2' (CO_{2eq}) that would produce the same amount of warming.

regulations led to reductions in aerosol emissions). Working Group I of the IPCC concluded in 2013 that human activities – the release of greenhouse gases and aerosols – were extremely likely to be the dominant cause of the observed warming since the mid-20th century.[9]

THE EFFECTS OF CHANGES IN THE ATMOSPHERE ON WEATHER AND CLIMATE

Greenhouse gases are well-mixed in the atmosphere, so affect the global energy budget. However, whilst everywhere can be expected to become warmer, the precise effects on weather and climate vary from place to place. This is because the climate of a place is determined not just by the local energy balance, but also by the circulation of the atmosphere and the redistribution of energy and water around the planet. Changes in the energy balance will affect weather and ocean circulation patterns, perhaps by changing their positions and perhaps by changing their intensities. Thus, whilst we would expect that a warmer world would generate more frequent intense rainstorms because a warmer atmosphere can hold more water, in a specific place this might be offset or exaggerated by changes in the occurrence of different weather patterns.

We can also expect to see changes in one part of the climate system feeding through to changes in another. For example, it has long been known that increases in temperature would lead to reductions in snow and ice cover in the Arctic, which would lead to further increases in temperature because the surface of the land and sea would be darker than when it was covered

9 IPCC (2013), Summary for Policymakers. In: *Climate Change 2013: The Physical Science Basis. Contribution of Working Group I to the Fifth Assessment Report of the Intergovernmental Panel on Climate Change*. Stocker, T., et al., eds. Cambridge University Press.

by snow and ice; more energy would therefore be absorbed rather than reflected back into the atmosphere. This is another example of a positive feedback, and high latitudes are expected to warm more rapidly than low latitudes. However, it is also now thought possible that the position of the jetstream, which drives circulation and weather patterns across the north Atlantic and much of Europe, might be influenced by temperature gradients across the Arctic. Whether or not the influence is very strong, this illustrates the potentially complicated chain of drivers that links increasing greenhouse gas concentrations to changes in weather at a place.

OTHER EFFECTS OF INCREASING GREENHOUSE GASES

An increasing concentration of greenhouse gases does not just affect weather and climate. As oceans become warmer, water expands, and as glaciers and ice sheets melt due to higher temperatures they release more water into the oceans. Over the 20th century, it is estimated that the global average sea level has risen by around 19 cm,[10] and the rate of rise is increasing. The rise in sea level has varied around the coastline, partly because different parts of the ocean have been rising at different rates, and partly because land at the coast in some places is rising or falling. These local changes in land level may be due to the slow recovery of land after the weight of ice was lifted after the end of the most recent ice age (as is the case in northern Europe, for example), or may be caused by over-extraction of water (as has happened under several large coastal cities in Asia).

10 IPCC (2013), Summary for Policymakers. In: *Climate Change 2013: The Physical Science Basis. Contribution of Working Group I to the Fifth Assessment Report of the Intergovernmental Panel on Climate Change*. Stocker, T., et al., eds. Cambridge University Press.

CO_2 fertilises plant growth, so we can expect that some types of plants will grow more vigorously with higher concentrations of CO_2 in the atmosphere. There is some evidence that tree growth has accelerated in some regions since the early 20th century, probably largely because of higher CO_2 levels. Different plants are differently sensitive to CO_2 concentration, and whilst some major agricultural crops would likely see increased growth, others would be less affected. Any effects of increases in CO_2 concentration, of course, would be combined with the effects of changes in weather and growing conditions. Some of the additional CO_2 emitted into the atmosphere is absorbed by the oceans (approximately 30% so far), which leads to ocean acidification potentially affecting sea life and ecosystems.

HAS CLIMATE CHANGE ALTERED RECENT WEATHER?

Over the last few decades, global mean surface temperature and global mean sea level have both increased, and average temperatures have increased virtually everywhere. At the same time, snow cover in spring in the Northern Hemisphere has declined, the extent of Arctic sea ice at the end of summer has reduced, and the upper levels of the ocean have become warmer.[11] There have been an increased number of warm days in most areas, and there have been more frequent heatwaves across many regions. The number of cold days has generally decreased. In many areas – where data are available – there have been more frequent heavy rainfall events.[12] All of these

11 IPCC (2013), Summary for Policymakers. In: *Climate Change 2013: The Physical Science Basis. Contribution of Working Group I to the Fifth Assessment Report of the Intergovernmental Panel on Climate Change*. Stocker, T., et al., eds. Cambridge University Press.

12 For example in the UK: Jenkins, G.J., et al. (2007), *The Climate of the United Kingdom and Recent Trends*. Met Office Hadley Centre, Exeter, UK.

large-scale trends have been attributed to the increasing concentration of greenhouse gases in the atmosphere, in the sense that trends can only be explained with reference to the rising concentrations. At any individual location it is difficult to quantify recent changes in weather statistics, because underlying climate change is superimposed onto local effects of internal climatic variability, but over the large scale it is clear that many dimensions of weather have changed in recent decades.

However, it is much more difficult to say whether individual extreme events, such as storms, floods and heatwaves, have become more or less frequent. There are a number of reasons for this. Extreme events are by definition rare, and the available records tend to be quite short. Often many factors are changing at the same time. The total economic losses from flood events (globally) are increasing, for example, but this is largely because the exposure to loss in flood-prone areas is increasing. Finally, 'extreme' events have always happened, and records would be broken even if there were no underlying change. An analogy is frequently drawn with rolling dice. Occasionally a pair of sixes is thrown – the chance is one in 36. Sometimes a second pair of sixes is thrown. This would be unlikely, but is not impossible. We would only begin to suspect that the dice were loaded if we began to get more pairs of sixes than we expected. In other words, one pair of sixes is not evidence that the dice are loaded. We could have thrown them with a pair of true dice. We can only tell if the dice are loaded after a number of throws, when we can see a trend.

A number of attempts have been made to see if the dice are loaded by comparing the likelihood of some recent extreme events under current climatic conditions, and under the conditions that would have pertained if there had been no climate change. This is, clearly, technically challenging, as we do not know precisely what would have happened to climate without the increasing forcing from greenhouse gases and the other drivers of climate change. Nevertheless, it has been shown that events similar to the European heatwave of 2003, the heavy autumn rainfall in the UK in 2000 and the high summer temperatures in the USA in 2012 are substantially more likely to have occurred in the current climate than without the observed change.[13] In other words, the dice (or at least some of them) are loaded.

However, some notable events are no more likely to have occurred now than in the past or in a parallel universe with no climate change. The low rainfall in the USA in 2012, for example, was no more likely to have happened under the current climate (even though the high temperatures that occurred at the same time were probably influenced by climate change), and similarly the high rainfall across eastern Australia in 2012 was not unusual. In both cases, the extreme rainfall was a manifestation of the year-to-year climatic variability caused by the internal workings of the climate system as discussed above.

Recent summers in the UK and northern Europe have been notably wetter than average, and indeed rainfall has been very volatile; the wet summer of 2012 in the UK followed a very dry winter. The recent wet summers and very wet winter of 2013/2014 have been attributed to a southward shift of the storm tracks that move across the Atlantic and bring rain to western and northern Europe. This has occurred because the high-level jetstream which drives the storms has been running further south than usual – but it is less clear why

13 Peterson, T.C., et al. (eds) (2013), Explaining extreme events of 2012 from a climate perspective. *Bulletin of the American Meteorological Society* 94, S1-S74.

the jetstream shifted position. The most likely explanation is that the North Atlantic Ocean has been warmer than average, and when this happens the jetstream moves further south; it is also possible that reductions in sea ice in the Arctic Ocean have caused the jetstream to shift southwards, but this is more speculative. But why is the North Atlantic Ocean going through a warm phase? It may be due to 'normal' internal variability within the climate system, but may be influenced by the increasing concentration of greenhouse gases in the atmosphere and increasing amounts of heat stored in the oceans – and therefore be a manifestation of the effects of climate change.

In sum, whilst it cannot be stated definitively that climate change has led to increases (or decreases) in the frequency of all extreme weather events, we can confidently assume that climate change has already altered many weather risks: it has already loaded the dice.

HOW MIGHT CLIMATE CHANGE IN THE FUTURE?

Our assessment of how climate might change in the future is beset by three uncertainties. First, we do not know precisely how emissions of greenhouse gases will change in the future, because we cannot predict economic development (or, more precisely, future energy use). This is inherently uncertain; more than a few years into the future, we can only make plausible assumptions – called scenarios – about future energy use. Second, even if we did know how emissions were to change, we cannot predict precisely how the climate system will respond. This is because although we know the basic physical principles underpinning climate change, how change happens depends on the detailed workings of the climate system. As our understanding of the climate system improves, so this uncertainty should reduce. Third, we have already seen that the climate system is characterised by internally induced variability. Any effect of underlying climate change will be superimposed on this variability, and may indeed change it, so it might be difficult to see a detectable climate change signal for a few years. Our projections of how climate might change over the next few years are therefore also dependent on our understanding of what causes this internal variability, and on our ability to observe the current climate system.

These three sources of uncertainty are important at different time scales. For the next couple of decades, the effects of different assumptions about future emissions are relatively unimportant because the climate system responds only slowly to changes in the energy balance. However, over this time-scale climatic variability will likely be large relative to the effects of an underlying increase in greenhouse gas concentrations, and what we project will happen will therefore be very much dependent on our understanding of causes of variability and where we start from (our 'initial conditions'). More than 30 years or so into the future, the climate change signal will be stronger and our projections are more influenced by our ability to simulate climate and how it will respond to forcing. After around 50 years, uncertainty in future emissions begins to affect our assessment of how climate will change. The rest of this section looks at these three sources of uncertainty in more detail.

The future emissions of greenhouse gases depend on how much energy is used, how that energy is produced, and how much land is cleared. This is partly dependent on future population levels, but is very strongly influenced by the characteristics of future economic growth. There are, of course, a great many plausible scenarios for future population and economic growth, and therefore future emissions. At present, just over 10 billion tonnes of carbon (equivalent to over 37 billion tonnes of CO_2) are emitted by the burning of fossil fuels and land degradation (Figure 2.3), nearly 50% more than in 1990. The International Energy Agency's World Energy Outlook 2012 projects increases of between 22% and 45% by 2035, under its Current and New Policy Scenarios.[14] Figure 2.5a shows four different plausible future patterns of emissions of carbon dioxide from the combustion of fossil fuels,[15] along with observed emissions to 2012.[16] The labels for the four scenarios are not relevant here – they represent the strength of the forcing these emissions have on the atmosphere – but there is clearly a large variation between them. Ignoring the lowest scenario (RCP2.6) for the moment, the potential emissions from the burning of fossil fuel in 2050 range from 11 to 20 billion tonnes of carbon per year (or 41 to 76 billion tonnes of CO_2 per year) – approximately 80% to 230% higher than in 1990. To this, we can add the contributions from other greenhouse gases such as methane and nitrous oxides, which again we cannot predict precisely so have to base on plausible scenarios (and note that the four scenarios shown in Figure 2.5 are a subset of a much wider range of potential future emissions scenarios).

The effects of increasing emissions of greenhouse gases on climate are estimated using computer simulation models. There is a range of types of model used for different applications, but all are based on the physics of the climate system. The

14 International Energy Agency (2012), *World Energy Outlook*.

15 Van Vuuren, D., et al. (2011), The representative concentration pathways: an overview. *Climatic Change* 109: 5-31.

16 Global Carbon Budget 2013, http://www.globalcarbonproject.org/carbonbudget/.

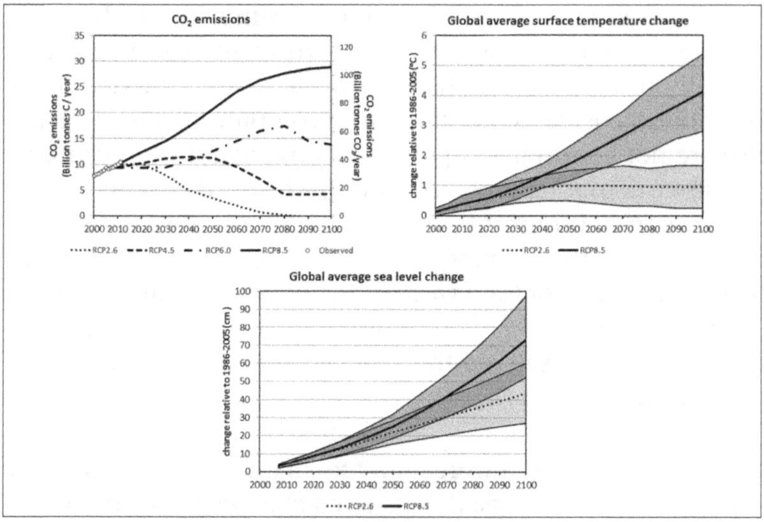

Figure 2.5 Potential change in CO_2 emissions, global average surface temperature and sea level rise. The observed emissions are taken from the Global Carbon Budget 2013. The projected changes in temperature and sea level are taken from the IPCC AR5 Working Group I Summary for Policymakers

most complicated type of model applies the laws of physics governing the movement of energy and water, and simulates the evolution of weather over time. Such models work on a gridded basis, dividing the ocean and atmosphere into a series of layers. The same type of computer model is used to make weather forecasts, predicting the movement of weather systems and such important aspects of weather as temperature, rainfall and windspeed. There are three main differences in practice between the models as used in weather forecasting and for simulating future climate. The weather forecast models simulate just a few days or weeks into the future, they are continually updated with weather observations, and they are typically run over a much finer grid than the models used for climate projections. The climate models simulate weather over decades or even centuries, and this is the main reason why they typically run at a coarser scale than the weather models; they are very expensive to run in terms of computer resources. They also are not continually updated by observations – because of course we do not have observations in the future – so in a sense they 'run free'. Whilst they simulate weather at the daily or finer time scales for the next few decades, we do not believe literally their predictions for (say) weather in London on July 17th 2052. Instead, we focus on their projection of average climate in the future – for example the average summer temperature in southern England in the 2050s.

The projections of future climate made by these models depend on three things: the assumed change in forcing of the climate, the way the models represent climate processes, and the starting point for the simulation (the 'initial conditions'). The first depends of course on the future emissions of greenhouse gases – which as we saw above we cannot know perfectly – but also includes such effects as changes in solar radiation which are more predictable. Different climate models represent processes in the climate system differently, although they are all fundamentally based on the same equations. They differ in the spatial resolution at which they apply the equations and in the approximation of processes that cannot be simulated explicitly. With the same forcing, and the same initial conditions, different climate models simulate broadly the same climate, but the regional and local details may be different. The same model run from different sets of initial conditions would also produce slightly different simulations each time, so sometimes a climate model is run several times from different starting positions to produce an *ensemble* of projections (this approach, incidentally, is also widely used in weather forecasting).

Figure 2.5b shows the projected change in global average surface temperature for two of the scenarios for future emissions shown in Figure 2.5a. The range in projected change for each emissions scenario reflects the range between estimates from different climate models. If emissions follow the high 'RCP8.5' scenario, the global average surface temperature would be 1.2-2.3°C higher than present (more precisely, the 1986-2005 mean[17]) around 2050, and between 3 and 5.4°C higher at the end of the 21st century. To place this in context, remember that the earth has warmed up by around 0.6°C (Figure 2.2) since the middle of the 19th century. The second projection in Figure 2.5b shows change in temperature under the low 'RCP2.6' scenario. Under this scenario, the increase in global average surface temperature is 'only' 0.3 to 1.8°C, with an average estimate of around 1°C. However, these two scenarios are not equally likely, and in fact were based on a different underlying philosophy. The RCP8.5 scenario assumes emissions continue to rise unchecked, whilst the RCP2.6 scenario assumes very substantial reductions in global emissions. As shown in Figure 2.5a, RCP2.6 assumes 'negative emissions' after 2070, meaning that more CO_2 is absorbed than is emitted. International climate policy and the targets for climate change agreed by the international community are discussed in Chapter 4, and we just need to note here that RCP2.6 is broadly consistent with the most stringent targets being considered. From Figure 2.5b we can see that there is little difference in temperature change between the two extreme scenarios until after 2030, and this is because the climate system responds slowly to a change in the atmosphere. We are already committed to a certain amount of climate change because of past emissions. A corollary, of course, is that it will take many years before the effects of reductions in emissions are felt.

17 The global average surface temperature over the period 1986-2005 was approximately 0.3°C warmer than over the period 1961-1990 as shown in Figure 2.2.

Whilst global average temperature is an immediately obvious indicator of climate change, it is not necessarily the most useful in practice. Few of the consequences of climate change are related directly to global average temperature change; they generally depend on local changes in temperature and other aspects of weather, or on change in sea level. Figure 2.5c shows the change in average sea level under the two emissions scenarios in Figure 2.5b. By 2050, average sea level would be 19-32cm higher than present under the high emissions scenario, and only slightly lower than this under the low scenario.

The global average temperature change hides a big contrast between land and sea. Land will warm up more rapidly than the oceans, so the temperature increase over land will be higher than the values shown in Figure 2.5b. Also, temperatures will increase more at high latitudes than at low latitudes. This is primarily because higher temperatures generally mean snow lasts for a shorter time. Bright snow reflects incoming energy more than dark soil, so as the period when the land surface is dark increases, the proportion of energy reflected back decreases and temperature rises further. Higher temperatures have a direct effect on local weather regimes. Simply by shifting the mean climate upwards, high temperature extremes would become more common and low extremes less common. Warm air can hold more water, so heavy rainfall events would become more frequent. However, changes in local rainfall will also be affected by changes in weather regimes or, more precisely, by a change in atmospheric circulation. As the earth warms – with different amounts of warming in different places – the circulation of the atmosphere is perturbed. This means, for example, that the frequency, position and intensity of the depressions that bring most rainfall to the UK could change. As a general rule, it is likely that wet regions (or seasons) would become wetter and dry regions (or seasons) drier as greenhouse gas concentrations increase. Monsoons – which bring summer rainfall to south and east Asia, northern Australia, west Africa and the south west United States – are projected to last for longer and to generate more rainfall. However, the *size* of the change in rainfall in a given place is difficult to predict precisely.

Current climate models project little change in the overall frequency of tropical cyclones (known as hurricanes in the Atlantic Ocean and typhoons in the Pacific Ocean), although it is likely that very extreme cyclones may occur more frequently. However, because sea level is rising, there is highly likely to be an increased frequency of coastal flooding from the storm surges associated with tropical cyclones: with a higher sea level, a smaller cyclone is needed to generate a storm surge of a given height. Storms also occur outside the tropics, and these events (known as extra-tropical cyclones) bring not only heavy rainfall but also strong winds. Like tropical cyclones, extra-tropical cyclones are difficult to simulate in global climate models because they are generated and work at very fine spatial scales. Nevertheless, current models suggest that extratropical storms across the North Atlantic may become more frequent, with storms extending further into Europe.

Earlier on, we introduced the concept of internal climatic variability, whereby patterns or rhythms within the climate system lead to periodic fluctuations in the climate regimes experienced at a place. At the global scale, the dominant pattern of variability is linked to El Niño/Southern Oscillation (ENSO), but in other regions other rhythms may be more important. The variability in weather across Europe in winter, for example, is linked with the North Atlantic Oscillation (NAO: a fluctuation in atmospheric pressure across the North Atlantic

Ocean). Variations in the temperature in the North Atlantic Ocean affect rainfall across Europe. Climate change due to increasing concentrations of greenhouse gases is superimposed onto this variability, and for the next couple of decades this variability may mask an emerging signal of climate change. This is particularly likely to be true for changes in rainfall and the occurrence of extreme storms, where year-to-year variability is high. Therefore, whilst the long-term trend in a place might be for rainfall to increase in winter and decrease in summer for example ('the wet get wetter and the dry get drier'), for a few years that place may see dry winters and wet summers because of natural internal climatic variability. The underlying change in temperature due to increased greenhouse gases is large compared to year-to-year variability, so we might expect to see a clear signal for temperature sooner than for rainfall.

FORECASTS, PREDICTIONS AND PROJECTIONS

The previous section has shown how our estimates of the future are influenced by a number of factors, some of which are more predictable than others, and which have different effects over different time-scales. It is therefore necessary to clarify here the distinction – in meteorological and climatological terms – between forecasts, predictions and projections.

A *forecast* is an estimate of weather in the near future, based on what we know about the current state of the atmosphere, ocean and land surface. Forecasts are frequently updated as more information becomes available, and the quality or skill of a forecast system can be evaluated quickly because we can rapidly compare what is forecast with what actually happens. Forecast systems can therefore in principle be continually improved, and we can quickly learn how to interpret forecasts and correct for bias.

A *prediction* is an estimate of weather over the longer term – perhaps up to a few months – based on what we know about the current state of the system and how we think the external driving 'boundary' conditions will change. These external boundary conditions will include the variation in the receipt of solar radiation through the year – which is relatively easy to predict – and variations over time in the state of the ocean. This may be possible to predict too, because oceans warm and cool relatively slowly. The key difference between a forecast and a prediction is that we cannot rapidly update a prediction with new information. Forecasts may be updated every few hours, but there is little point in updating predictions so frequently. The quality of predictions can be tested and prediction systems improved, but only by testing against past experience.

A *projection* is an estimate over the much longer term, conditional upon some assumptions – or scenarios – of how unpredictable forcings will change over time. Emissions of greenhouse gases are, for example, inherently unpredictable, so we have to make assumptions about how they will change before running our models to simulate future climate. The estimates of future climate are therefore conditional on these assumptions. We also know that our climate models are relatively simplistic representations of a complex system, and become less reliable once they move further away from observations. Projections are therefore best thought of as plausible future climates.

Forecasts, predictions and projections can all be made with the same type of climate model, which simulates weather at the very fine time-scale – typically just a few hours. However, the types of information that we can plausibly extract from these models differs. We can use the output from a forecast model to say, for example, that the rain that is likely to fall tomorrow will fall in the afternoon, or that Sunday is more likely to be wet than Monday. When we use a similar model in prediction mode, however, we are less confident about estimates for a particular day, but we are more confident in saying that, for example, the temperature in two months time is likely to be above or below the average. When we use a similar model to make projections, we have greater confidence in our estimates of future climatology, rather than weather in specific months or years.

CONCLUSION

This chapter has outlined the basic physics behind climate change, and shown how climate may change in the future. The key points are:

- The basic physics explaining the effect of increasing emissions of greenhouse gases on the atmosphere is well established, and the fundamental principles have been known for over 150 years.

- The earth's climate is showing evidence of change, particularly in terms of surface temperature and sea level, and this is largely due to human emissions of greenhouse gases.

- We can no longer assume that past experience provides a good guide to the likelihood of the occurrence of extreme weather events.

- Our future climate depends on future emissions, but for the next few decades the climate will still be slowly reacting to our past emissions and variations from year to year will be strongly influenced by internal rhythms and variability within the climate system.

- We can project the broad patterns of change in temperature and rainfall across the globe, using climate models, but precise quantitative projections at a place are not possible. Our projections of local change are therefore uncertain – and this uncertainty is extremely important.

- Policies to reduce emissions will take a few decades to have an effect on global climate, but the sooner emissions begin to reduce the greater the long-term effect.

- Climate change is superimposed onto natural variability in the climate system, and for the next couple of decades this might hide or exaggerate an underlying climate change.

3 The Consequences of Climate Change

INTRODUCTION

Organisations are exposed to risk from 'extreme' weather events, such as floods, windstorms, heatwaves, cold spells and droughts. The occurrence of an event may have consequences for an organisation, so it will often put in place measures to minimise their impacts. This may include some form of physical protection, the purchase of insurance or the preparation of contingency plans. An organisation may decide to accept a risk, or indeed be unaware of it. Organisations are also – usually implicitly – tuned to the current climate. The type and design of their product or service, the layout of their premises and their assumptions about the availability of resources and raw materials are all influenced by current climate and its variability. With climate change, however, we can no longer assume that the past is a good guide to the future. The climate that we are tuned to has changed, and will change more in the future.

This chapter summarises the potential consequences of climate change, first by area of impact and then from the perspective of an organisation. In practice, organisations may adapt to a changing climate, in the same way that they have adapted to current climate and the chance of experiencing an extreme event. The last section of the chapter discusses some of the barriers to adaptation to climate change.

THE IMPACTS OF CLIMATE CHANGE

In Chapter 2 we saw that the increasing concentration of greenhouse gases has led to increases since the middle of the 20th century in temperature and sea level, and reductions in snow and ice cover. There is also increasing evidence that these changes have also begun to impact upon many natural systems.[1] Many terrestrial, freshwater and marine species have shifted their geographic ranges, migration patterns and abundance in response to changes in temperature in particular. Higher temperatures have led to reductions over time in wheat and maize yields in many regions. Impacts so far on human systems, however, are much more difficult to identify because many human systems have undergone substantial change in themselves over time; it is difficult to see a signal of climate change.

Chapter 2 described the future large-scale and global effects on climate and sea level of an increasing concentration of greenhouse gases. The impacts of future climate change on natural and human systems, however, are determined by how weather regimes and sea level change at the local

1 IPCC (2014), Summary for Policymakers. In: *Climate Change 2014: Impacts, Adaptation and Vulnerability. Contribution of Working Group II to the Fifth Assessment Report of the Intergovernmental Panel on Climate Change*. Field, C., et al., eds. Cambridge University Press.

scale, and are typically estimated in a two-stage process. The first stage is to construct *scenarios* for change in local climate or sea level from the global-scale simulations made with climate models run with increasing concentrations of greenhouse gases.[2] There are many ways of doing this, and the academic literature is full of examples. The details of these methods are not important here, but suffice to say that this 'downscaling' to the local scale adds an extra level of uncertainty to projections of local climate to the uncertainties introduced in Chapter 2. The second stage uses another model to estimate the consequences of a change in climate or sea level for some variable or indicator of interest, such as crop productivity, risk of flooding or energy requirements. This too adds another level of uncertainty, but this varies considerably from sector to sector depending on how well established the models are and how complicated the sector is.

The impacts of climate change are expressed relative to 'baseline' conditions in which there is no climate change. This baseline period varies between studies. Many studies use the 30-year period 1961-1990 to represent 'current' climate (30 years has long been assumed by climatologists to be sufficient to characterise the climate regime of a place), but increasingly studies use later periods such as 1971-2000 or 1981-2010. Estimated impacts are different with different baselines. This is partly because successive baselines have a higher average temperature (globally the period 1981-2010 was approximately 0.3°C warmer than 1961-1990, for example), and partly because the mean climate at a place varies from one 30-year period to another due to natural internal variability.

2 Note that there are other ways of constructing scenarios which can be used for sensitivity studies: these are discussed in more detail in Chapter 5.

The impacts of a changing climate depend not only on the change in weather or sea level, but also on the characteristics of the system being impacted. In some circumstances, for example, an increase in temperature of 2°C may have no real consequences, whilst in others a rise of just 0.5°C could have costly effects. It is therefore difficult to extrapolate results from one impact study to another situation. This is illustrated in Figure 3.1, which shows how the impact of climate change depends on the interaction between the change in climate, exposure and vulnerability. *Exposure* represents the presence of people and assets in places that could be affected by climate

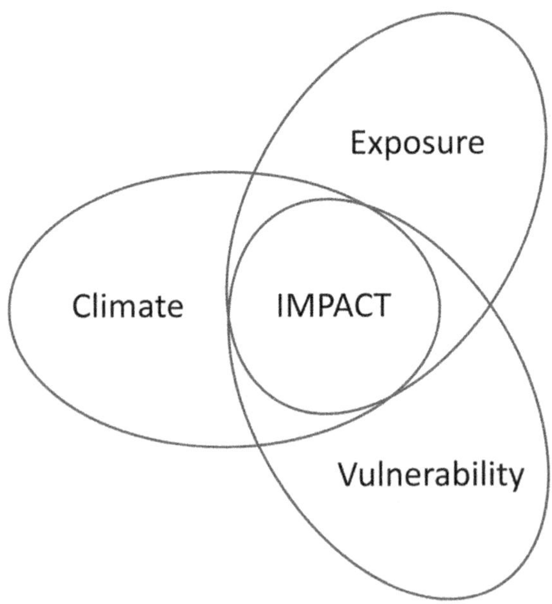

Figure 3.1 **Impact as a function of climate change, exposure and vulnerability (revised from IPCC (2011) Summary for Policymakers, *Special Report on Managing the Risk of Extreme Events and Disasters to Advance Climate Change Adaptation (SREX)*)**

change, such as on a river's floodplain. *Vulnerability* represents the propensity of those people or assets to suffer adverse consequences; this is dependent on the ability to absorb and recover from loss.

Most studies of the impact of climate change assume that no measures are taken to alleviate any impacts, or in other words that there is no *adaptation* to climate change. The studies therefore usually represent a 'worst case' assessment in the absence of any response, and indeed most are designed to see whether or not it is necessary to adapt explicitly. In practice the impacts of climate change will likely be less than implied by published impact studies because organisations and individuals will adapt, but it is currently very difficult to estimate how effective (and costly) this adaptation will be.

CHANGES IN CLIMATE AT A PLACE

Before examining the potential impacts of climate change across a range of impact areas, it is worth illustrating in more detail how climate change may alter the weather of a place.

In the UK, the UKCP09 climate projections[3] present a coherent set of plausible representations of possible future climate. They show changes in climate across the UK in probabilistic terms, and are based on the analysis and interpretation of many different climate models (see Chapter 5). Projections are made for low, medium and high rates of future emissions of greenhouse gases. The headline messages from the UKCP09, under the medium emissions assumption are:

3 Murphy, J., et al. (2009), *UK Climate Projections Science Report: Climate Change Projections*. Met Office Hadley Centre, Exeter. ukclimateprojections.metoffice.gov.uk.

- By 2050, average winter temperatures would be likely to be between 1.1 and 3.4°C higher than over the period 1961-1990 in southern England, and 1.1-3.1°C higher in the north. Average summer temperatures would be 1.3 to 4.6°C higher in the south, and 1.2 to 4.1°C higher in the north.

- By 2050, summer maximum temperatures are likely to be 1.3-4.6°C higher in southern England and 1 to 5.7°C higher in the north.

- In southern England, average winter rainfall is likely to be 2-36% higher, and summer rainfall will be between 41% lower and up to 7% higher. In the north, average winter rainfall is likely to be 1 to 24% higher, and summer rainfall between 30% lower and just 1% higher.

- Average windspeed would likely be slightly lower (by at most around 5%) across the UK, but the effect is small.

- There will be little change in the frequency or intensity of windstorms across the UK.

- The frequency of fog is likely to be reduced across the UK, except in southern Britain in winter.

- By 2050, sea level in the Thames estuary and around the south coast would be between 11 and 33cm higher than in 1990; the rise would be lower further north because the land in southern England is falling whilst northern UK rises.

These changes relate to average climate, and some years will of course continue to be very different to the average. We can expect, therefore, to continue to experience occasional dry or cold winters and wet or cool summers, even though – on average – winters will be warmer and wetter than now and summers warmer and drier.

Figure 3.2 shows the potential change through the 21st century in average winter and summer temperature and rainfall in Reading, Berkshire, assuming a medium level of emissions. The ranges represent the uncertainty in projected changes, and whilst they are clearly large, the general direction of change is clear. Temperatures will increase more in summer than in winter, and rainfall is highly likely to increase in winter and very likely to decrease in summer. Current and 2050 average

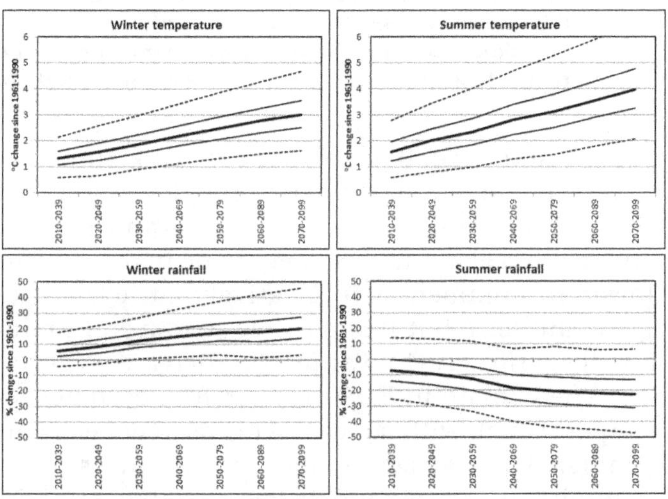

Figure 3.2 Change in average winter and summer temperature and rainfall in Reading, UK: UKCP09 climate projections, with a medium emissions pathway. The thick lines show the median change, the dotted lines show the 10 and 90% likelihood range, and the thin lines show the 33 to 67% likelihood range

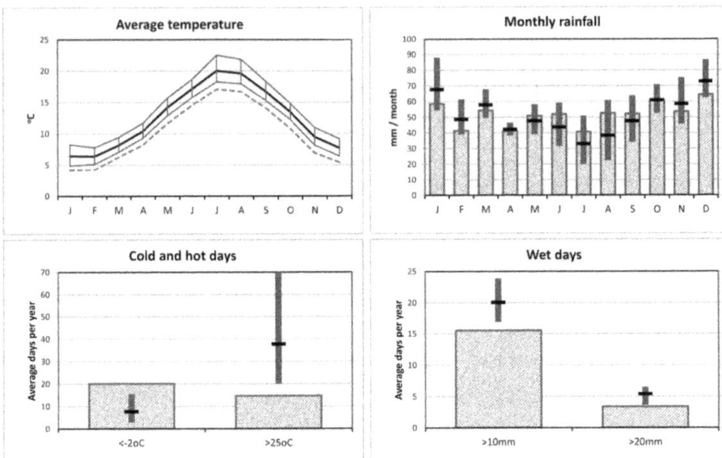

Figure 3.3 **Current and future (2050) weather in Reading, UK. Climate projections from UKCP09, with a medium emissions pathway. The future weather is shown as a range, based on the UKCP09 10 to 90% likelihood range**

temperature and rainfall are also shown in Figure 3.3. By 2050, average temperature is between 0.7 and 4°C higher than at present in winter, and between 1.2 and 5°C higher in summer. Average rainfall is likely to be higher in winter and lower in summer, with relatively little change in spring and autumn, but it is possible that average rainfall could reduce slightly in winter or increase slightly in summer. The climate in Reading therefore becomes more obviously seasonal, with a greater contrast between winter and summer. The average number of 'cold' days per year (with a minimum temperature below -2°C) falls from 20 to between 2 and 15 days, but the average number of 'hot' days (maximum temperature above 25°C) rises from 15 to between 20 and 70 days per year. The number of 'wet' days increases too, with perhaps a doubling of the number of days that rainfall is greater than 20mm.

Climate change, of course, varies from place to place, and Figure 3.4 gives an indication of the range in changes that could occur across the globe by 2050. The future climates in Figure 3.4 are estimated from a number of different climate models driven by the highest emissions scenario introduced in Chapter 2. In all cases, average temperature increases in each month; increases are slightly higher in high latitudes than in tropical regions. Changes in rainfall, however, are much more variable from place to place, and there is also greater uncertainty in projected changes at a place than for temperature. In some places and seasons – for example both western Russia and northeastern USA in winter and spring, and in northern India and eastern China in summer – there is a broadly consistent increase in average rainfall. In some regions – including Spain – there is a broadly consistent decrease in rainfall. But elsewhere, some climate models project an increase in rainfall whilst others project a decrease. This uncertainty makes it challenging to estimate the future impacts of climate change and, as we shall see in Chapter 5, is fundamentally important when considering how to react to the potential threats and opportunities from climate change.

Figures 3.2, 3.3 and 3.4 show the consequences of a change in 'average' climate, and a simple shift in the mean can lead to changes in the frequency of extreme events; this is effectively what is shown in the bottom part of Figure 3.3. Climate change also has the potential to change the variability in events from year to year, and also the shape of the distribution of events from one year to another. This is illustrated in Figure 3.5, which shows how a change in the frequency of extreme events (in this example hot days) can arise due to changes in mean, variability and distribution of extreme events. Under all three types of change, the frequency of hot days increases, but the frequency of cold days decreases, increases or remains

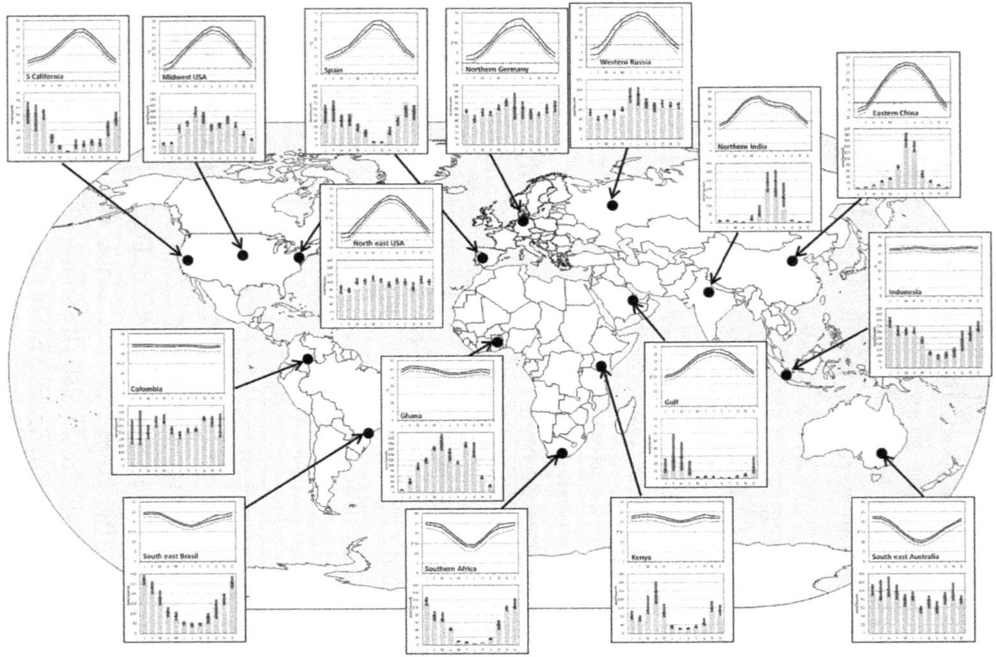

Figure 3.4 **Current and potential future (2050) average monthly temperature and rainfall for selected locations. The future climates (dark) are estimated from a number of different climate model simulations**

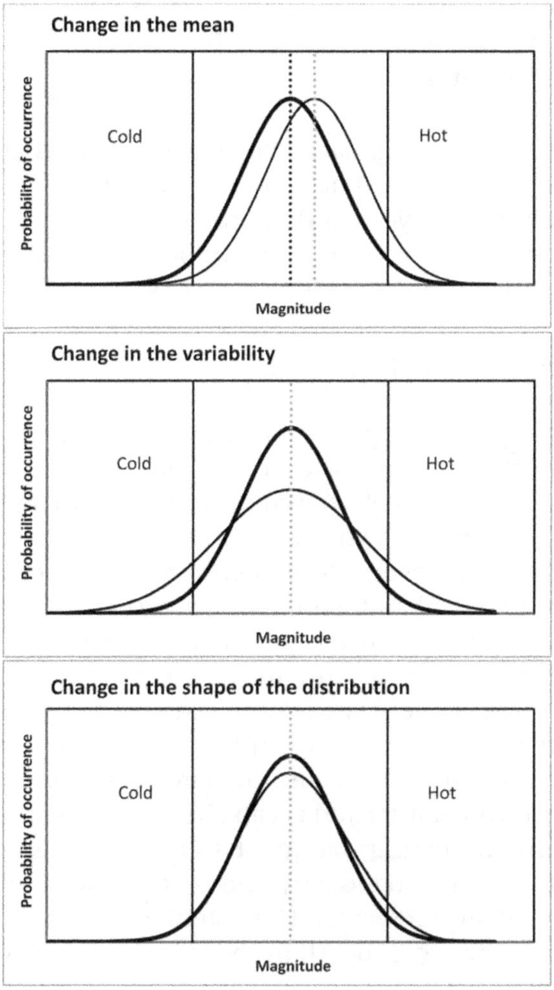

Figure 3.5 **The effects of changes in the mean, variability and symmetry of a frequency distribution on the occurrence of extreme events (from IPCC (2011) Summary for Policymakers, *Special Report on Managing the Risk of Extreme Events and Disasters to Advance Climate Change Adaptation (SREX)*)**

unchanged depending on whether the mean, variability or shape of the distribution changes. In practice, all three may change in the future.

At a place, therefore, climate change will lead to changes not only in average climate but also in the variability of weather events from year to year, and the frequency of occurrence of extreme events. Projected changes in both the mean and the variability can be very uncertain.

POTENTIAL IMPACTS BY SECTOR

There are many possible conceivable impacts of climate change on all sectors of the economy and society affected by weather, and the link between climate change and impact may be rather complicated. Figure 3.6 illustrates this schematically. The bottom of the figure shows some broad areas of impact – hazards, water, food and so on – and the top part summarises the sequence of links between climate change and impact. CO_2 and the other greenhouse gases alter the radiative balance of the planet (as explained in Chapter 2), and this changes temperature, other weather variables, the occurrence of extreme weather events and the circulation of the ocean. Changes in temperature and ocean circulation affect sea level. Changes in the concentration of CO_2 have direct impacts by affecting plant productivity and ocean acidity, which cut across several impact sectors. The potential impacts of climate change are reviewed in depth in the reports of the IPCC.[4]

Changes in physical hazards, such as floods, droughts and heatwaves, typically depend on changes in the frequency

4 IPCC (2014), Summary for Policymakers. In: *Climate Change 2014: Impacts, Adaptation and Vulnerability. Contribution of Working Group II to the Fifth Assessment Report of the Intergovernmental Panel on Climate Change*. Field, C., et al., eds. Cambridge University Press.

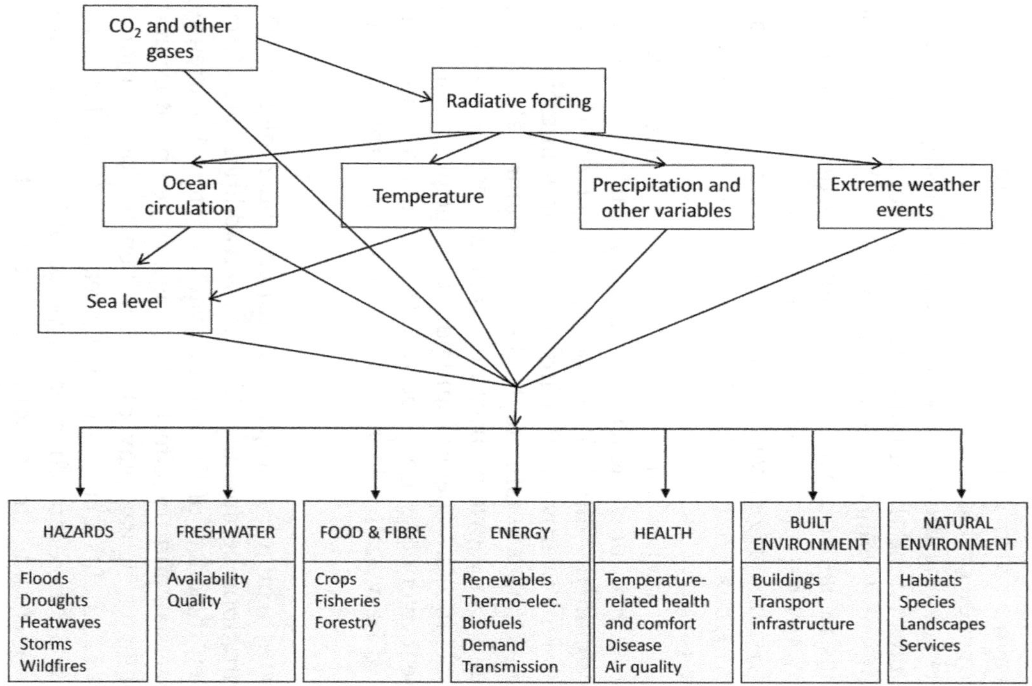

Figure 3.6 The links between climate change and impact

of extreme weather events. A gradual underlying change, however, may also gradually alter exposure to loss even if the frequency of 'extreme' events does not change. A good example is along the coast, where gradually rising sea levels would mean that the number of coastal floods (e.g. from hurricanes and typhoons) would increase even if the number of storms remained unchanged. Changes in the availability of freshwater depend on how river flows, groundwater levels and reservoir storage levels change. These are determined by changes in the amount and timing of changes in rainfall, which may be offset or exaggerated by the increased evaporation triggered by higher temperatures. The effects of change in rainfall on river flows depend on the physical characteristics of the catchment – especially its geology – so it is quite possible that different nearby catchments may respond differently to the same change in climate. Higher temperatures also affect the quality of river water (because it changes the rate at which chemical processes operate), and pollution incidents may also increase if river flows are reduced so there is less dilution or if increased heavy rainfall events flush pollutants from land to river.

The growth of most crops is increased under higher CO_2 concentrations, because the CO_2 acts as an additional fertiliser. However, the effect of this on the *quality* of crops is currently controversial, and the effects of changes in temperature and rainfall in particular may offset the benefits of additional CO_2; the benefits may also not be realised if nutrients are limiting or there are changes in pests and diseases. Some crops – including maize, sorghum and millet – fix carbon from the atmosphere using a different mechanism, and these do not see increased growth with higher CO_2 concentrations (although they may become more tolerant of drought). On the positive side, higher temperatures may mean that crops can be grown further north (poleward) than currently. Higher CO_2 concentrations mean most tree species – especially those used in commercial forestry – grow more vigorously, although as with crops, this effect may be offset by increased drought or storminess. Northern forests are likely to benefit not only from the higher CO_2, but also from longer growing seasons. Both ocean and coastal fisheries will be affected by increases in water temperature, changes in ocean currents and the increased acidity of sea water; the higher acidity directly reduces the ability of shellfish and corals to form their shells, and because plankton are also affected, influences the entire aquatic food chain.

Climate change directly affects the potential performance of renewable energy sources. Changes in river flows affect hydropower production, changes in windspeed may potentially affect wind power production, and changes in the amount of sunshine (due to changes in cloud cover) could affect solar power productivity (and higher temperatures would also affect the efficiency of photovoltaic cells). The productivity of biofuels is affected by climate change in the same way as other agricultural crops. The capacity of thermo-electric and nuclear power stations is also potentially affected by climate change, because higher air temperatures reduce efficiencies of cooling; higher water temperatures and lower flows would also reduce the cooling efficiency of power stations reliant on river water for cooling. The efficiency of electricity transmission networks reduces with increases in temperature, and changes to the frequency of ice and wind storms could alter the frequency of power outages. As ever, the precise impacts depend on local designs and circumstances.

Some components of the demand for energy are also related to weather and climate. In general terms, we would expect demand for energy for heating to reduce, but demands for cooling energy for air conditioning to increase. The effect of these changes on total energy demand would of course depend on the relative importance of heating and cooling demands, and how they compare with other demands.

Higher temperatures increase the risk of death from cardiovascular and respiratory ill-health amongst vulnerable people in hot summers, and also increase thermal discomfort and reduce labour productivity. On the other hand, a reduction in the occurrence of cold spells would likely reduce excess mortality – but cold spells typically make a small contribution to overall winter mortality. The spread of some major diseases involving insect vectors (such as malaria) may potentially be influenced by changes in climate which affect the geographic or seasonal distribution of insects, but the spread also often depends on the availability of suitable local habitats. Poor air quality contributes to ill-health, particularly in urban environments, and changes in temperature and windspeed in particular could alter the concentration and dispersion of pollutants.

We have already seen how higher temperatures would probably affect buildings and how we work within them; we would probably spend less energy on heating, but need to spend more on cooling in order to keep temperatures down to acceptable (or legal) levels. Individual buildings too may be affected by more intense rainfall, perhaps leading to increased failure of drainage systems. More widely within urban areas, more frequent intense rainfall has the potential to lead to increased urban flooding and associated disruption of transport systems. Transport networks are also affected by other extreme events, such as windstorms, snowstorms, icing and fog. We can expect the frequency of snowstorms and icing generally to decrease – but not disappear completely – but possible changes in extreme windstorms and fog are more uncertain. High temperatures also directly affect transport infrastructure. In very high temperatures, rails buckle and roads melt, and low temperatures contribute to the development of potholes in roads.

The natural environment of a place – its plants and animals – is closely controlled by climate, which affects what grows and therefore what is able to feed, and the timing of biological events such as flowering and the hatching of eggs. The potential effects of climate change on plants and animals (and their pests and diseases) are complex, and vary from place to place. The greatest impacts are likely to be on species that are at the limits of their range, where a change in climate would either make things considerably worse for them or substantially better. Whether changes in individual species have a substantial effect on entire habitats or indeed landscapes depends on how important the affected species are. The most vulnerable habitats and landscapes are likely to be in the highlands, in wetlands and along the coast.

Some examples of the potential impact of climate change by sector are given in Table 3.1. All are based on published case studies, and all assume that no adaptation takes place to reduce impacts. They make different assumptions about the rate of increase in greenhouse gas concentrations, and apply to different points in the future, but they do illustrate some of the potential consequences of climate change.

The potential impacts of climate change in a place depend strongly on local circumstances, and the effects of other changes that may be taking place such as altered demands for a service or land use changes. Our estimates of the potential

Table 3.1 Some examples of the impacts of climate change; all assume no adaptation to climate change

	Potential impact	Why?	Reference
HAZARDS			
Floods	Average annual losses from river flooding in the European Union increase by 2.2 to 3.4 times by 2080s, due to increased flood frequency	High river flows in many regions increase because of extra rainfall, but in some areas lower snowfall means spring peaks are lower	Feyen, et al. (2012), *Climatic Change* 112: 47-62
	By 2050, an extra 0.8 to 4 million people would be flooded each year in coastal floods	Rising sea level increases the chance of flooding occurring in the future	Arnell, N.W., et al. (2013), *Nature Climate Change* 3: 512-519
Storms	By 2050, average annual windstorm losses in Germany increase by 3-16%	Future windstorms are more intense, with higher maximum wind speeds	Donat, M.G., et al. (2011), *Natural Hazards and Earth System Sciences* 11: 1351-1370
FRESHWATER			
River flows	Summer river flows in southern England decrease by up to 80% by 2050	Rainfall in spring is reduced, and evaporation increased	Prudhomme, C., et al. (2012), *Hydrological Processes* 26: 1115-1118
FOOD AND FIBRE			
Crop productivity	Wheat yield in southern Africa decreases by approximately 17% after 2050	The effects of increased CO_2 concentrations on improved productivity are offset by higher temperatures and less rain	Knox, J., et al. (2011), *Environmental Research Letters* 7
Fisheries productivity	Productivity of coral reef fisheries may decrease by 20% by 2050	Higher water temperatures and increased acidity, together with changes in ocean circulation, reduce the amount of food available for fish	Bell, J.D., et al. (2013), *Nature Climate Change* 3: 591-599

Forest productivity	Timber yields from Norway Spruce in northern Finland increase by up to 68% by 2050	Increased CO_2 concentration increases tree growth	Kellomaki, S., et al. (2008), *Phil. Transactions Royal Society B* 363: 2341-2351
ENERGY			
Hydropower generation	Reduction in power generation potential by 2050 of approximately 8% in two Californian hydropower systems	Reductions in river flows lead to reductions in generated power	Vicuna, S., et al. (2011), *Climatic Change* 109: 151-169
Cooling energy demand	Increase in commercial cooling energy demand by 2-5% in Australian offices by 2030	Increased frequency of warm spells leads to increased use of air conditioning	Guan, L. (2012), *Building and Environment* 55: 8-19
System performance	California's electricity generation and distribution system would need an extra 11.5% peak capacity by the 2080s to provide current levels of demand	Higher temperatures reduce the capacity of power plants and the efficiency of transmission along the distribution network	Sathaye, J.A., et al. (2013), *Global Environmental Change* 23:499-511
HEALTH			
Heat-related mortality	Heat-related mortality amongst over 65s would increase in Californian cities by between 2 and 7.5 times by 2090	More frequent warm spells result in greater mortality from cardiovascular and respiratory illness	Sheridan, S.C., et al. (2012), *Climatic Change* 115: 311-326
BUILT ENVIRONMENT			
Rail network	Total cost of delays in hot summers in the UK doubles to £23 million by 2050	Higher temperatures increase buckling of railway track	Dobney, K., et al. (2010), *Proc. Inst. Mechanical Engineers Part F* 224: 25-34
NATURAL ENVIRONMENT			
Wildfire	The median number of summer moorland wildfires in the Peak District of England increases from one to 15 by 2050	Drier conditions and higher temperatures increase sensitivity to ignition	Albertson, K., et al. (2010), *Climate Research* 45: 105-118

impacts can also be highly uncertain, for two main reasons. First, there is uncertainty in how weather will change at a place, and particularly in how the frequency and characteristics of the extreme events that are often most critical to a system will change. This is well illustrated in Figures 3.2 and 3.3, where there is clearly a considerable range in possible future climates in Reading, Berkshire. Part of this uncertainty is caused by not knowing exactly how greenhouse gas emissions will change over the next few decades, but at the local scale the biggest contribution to uncertainty lies in our projections of detailed regional weather and climate.

Second, it is often difficult to estimate the effects of a given change in weather on a particular system. This can be because we do not precisely understand how the system works, because the response depends very much on the complicated interplay of different processes and local circumstances, or because we cannot model it precisely. This obviously varies from system to system. In some cases – such as in estimating the heating or cooling demands of a building for example – we have relatively well established and robust models which can be used to estimate the consequences of change, so the main uncertainty is in how weather regimes may change. In many parts of the natural world, however, projections of how a system will respond to change in weather are more uncertain. This uncertainty is important – and it is what makes adapting to climate change particularly challenging.

NATIONAL AND INTERNATIONAL ASSESSMENTS OF THE IMPACTS OF CLIMATE CHANGE

In the previous section, we saw how climate change has the potential to impact upon many different parts of the economy, society and natural world. Impacts will depend on local circumstances, and some will be much more 'important' than others. The perceived relative 'importance' of different impacts may also depend on priorities and values, with an obvious clear distinction between those with a focus on economic impacts and those more concerned with consequences for the natural environment.

The vast majority of studies into the potential impacts of climate change have focused on one particular issue or place, typically one that is believed by those conducting or commissioning the work to be of particular concern. It can be difficult to extrapolate from these studies – because they have not necessarily been designed to be representative of a broader picture – and in any case different studies tend to use different methods and scenarios. In order to get an appreciation of the relative importance of different impacts of climate change on a place or activity (or indeed an organisation), it is necessary to systematically identify all potential impacts and assess them consistently and systematically using the same set of assumptions and scenarios. Working Group II of the IPCC produces assessments of the impacts of climate change by sector and by major global region,[5] but these are largely based on expert interpretation of a wide range of studies that have used different approaches and scenarios.

A number of countries have conducted national assessments of the potential impacts of climate change. Assessments have been produced, for example, for the United States (the Third National Assessment was published in spring 2014[6]), Canada

5 IPCC (2014), Summary for Policymakers. In: *Climate Change 2014: Impacts, Adaptation and Vulnerability. Contribution of Working Group II to the Fifth Assessment Report of the Intergovernmental Panel on Climate Change*. Field, C., et al., eds. Cambridge University Press.
6 Melilo, J.M., Richmond, T.C. and Yohe, G.W. (eds) (2014), *Climate Change Impacts in the United States: the Third National Climate Assessment*. US Global Change Research Program.

(published in 2008)[7] and the Netherlands (updated in 2012).[8] The European Environment Agency published an assessment of potential impacts at the European scale in 2012[9] (Figure 3.7) and, at the finer scale, California published its third assessment in 2012.[10] Most such studies have been based on a compilation of existing published research together with

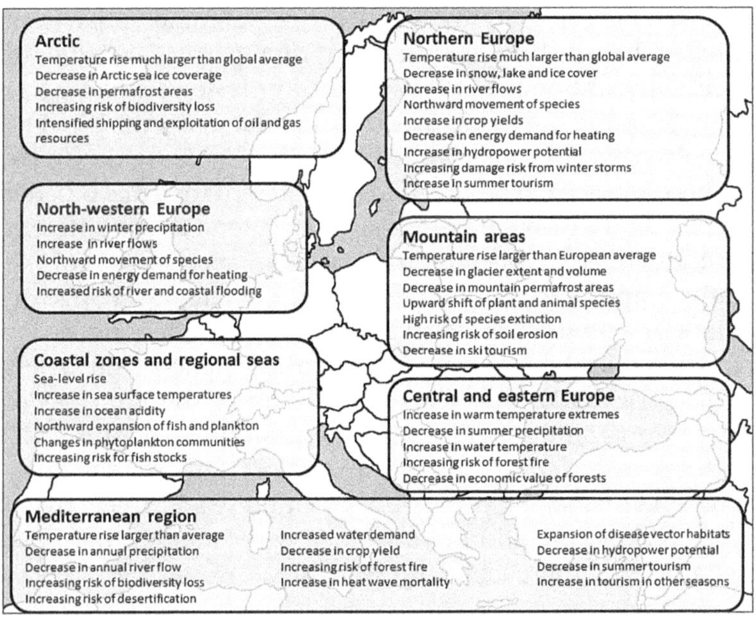

Figure 3.7 An overview of the potential impacts of climate change in the European Union (European Environment Agency, 2012)

7 Lemmen, D.S., et al. (2008), *From Impacts to Adaptation: Canada in a Changing Climate 2007*. Government of Canada, Ottawa.

8 Ligtvoet, W., et al. (2013), *The Effects of Climate Change in the Netherlands: 2012*. PBL Netherlands Environmental Assessment Agency.

9 European Environment Agency (2012), *Climate Change, Impacts and Vulnerability in Europe 2012*. EEA Report 12/2012.

10 Moser, S., et al. (2012), *Our Changing Climate 2012: Vulnerability and Adaptation to the Increasing Risks from Climate Change in California*. A Summary Report on the Third Assessment from the California Climate Change Center.

some new analyses conducted using a consistent set of climate scenarios. Many countries have included overviews of their exposure to the impacts of climate change in their national communications to the UNFCCC[11] (see Chapter 4), but these vary considerably in their detail.

The UK's Climate Change Risk Assessment (CCRA) was published in 2012.[12] Through consultation with a large number of organisations (public and private sector) around 700 potential consequences of climate change in the UK were identified, and these were whittled down to around 100 priority topics. Subsequently, the potential impacts for these priority topics were quantified using – as far as was feasible – the same set of climate scenarios, and the estimated impacts were categorised as having low, medium or high consequences, from either a social, environmental or economic perspective. The level of confidence in the estimated impacts was also categorised as low, medium or high. In general, confidence was assessed to be high where impacts were simply related to measures of average climate, and low either where impacts depended on changes in extreme weather events or where there were uncertainties in the precise relationship between change in climate and impact. The CCRA also separately identified threats and opportunities.

Table 3.2 shows the CCRA's overall summary. Whilst there is clearly a wide range in scope of the individual indicators – some are very broad (e.g. 'energy demand for cooling'), others are very specific (e.g. 'forest extent exposed to red needle blight'), and some impact areas (especially flooding) seem to be represented by several indicators – the CCRA provides a

11 unfccc.int/national_reports/items/1408.php.

12 Defra (2012), *UK Climate Change Risk Assessment: Government Report*. HM Government. Defra (2012), *UK Climate Change Risk Assessment: Evidence Report*. HM Government.

Table 3.2 Impacts of climate change in the UK, as assessed in the 2012 CCRA

	Timing			Confidence
	2020s	2050s	2080s	
Opportunities				
Decline in winter mortality due to higher temperatures				
Reduction in energy demand for heating				
Changes in wheat yield (due to warmer conditions)				
Opening of Arctic shipping routes				
Expansion of tourist destinations in the UK				
Opportunities to grow new crops				
Changes in fish catch (plaice/sole)				
Changes in grassland productivity				
Increase of potential yield of Sitka spruce in Scotland				
Threats				
Expected annual damage to residential property from flooding				
Ability to obtain flood insurance for residential properties				
Effects of floods and storms on mental health				
Insurance industry exposure to UK flood risks				
Residential properties at significant risk of flooding				
Summer mortality due to higher temperatures				
Forest extent affected by red band needle blight				
Overheating of buildings				
Energy demand for cooling				
Changes in species migration patterns				
Species unable to track changing climate space				
Number of unsustainable water abstractions				
Decline in marine water quality from sewer overflows				
Northward spread of invasive non-native species				
Risks to species and habitats from coastal evolution				

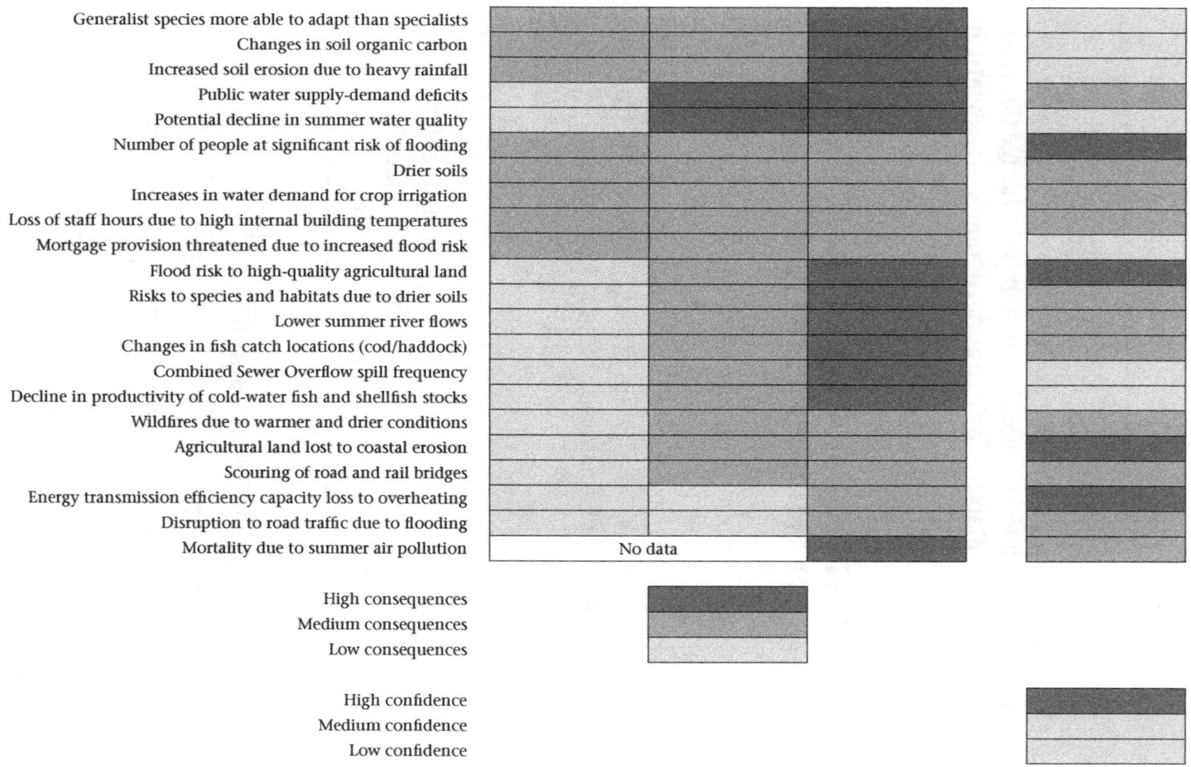
Generalist species more able to adapt than specialists
Changes in soil organic carbon
Increased soil erosion due to heavy rainfall
Public water supply-demand deficits
Potential decline in summer water quality
Number of people at significant risk of flooding
Drier soils
Increases in water demand for crop irrigation
Loss of staff hours due to high internal building temperatures
Mortgage provision threatened due to increased flood risk
Flood risk to high-quality agricultural land
Risks to species and habitats due to drier soils
Lower summer river flows
Changes in fish catch locations (cod/haddock)
Combined Sewer Overflow spill frequency
Decline in productivity of cold-water fish and shellfish stocks
Wildfires due to warmer and drier conditions
Agricultural land lost to coastal erosion
Scouring of road and rail bridges
Energy transmission efficiency capacity loss to overheating
Disruption to road traffic due to flooding
Mortality due to summer air pollution
No data
High consequences
Medium consequences
Low consequences
High confidence
Medium confidence
Low confidence

good first approximation at a comprehensive comparison of the effects of climate change. It highlighted changes in flood risk and water availability as being particularly significant in the UK.

THE POTENTIAL IMPACTS OF CLIMATE CHANGE ON AN ORGANISATION

In the previous section we looked at the potential impacts of climate change by sector – water, energy, food etc – which is the traditional way of considering climate change. However, it is possible to consider climate change differently, from the perspective of an organisation. Figure 3.8 represents such a perspective.

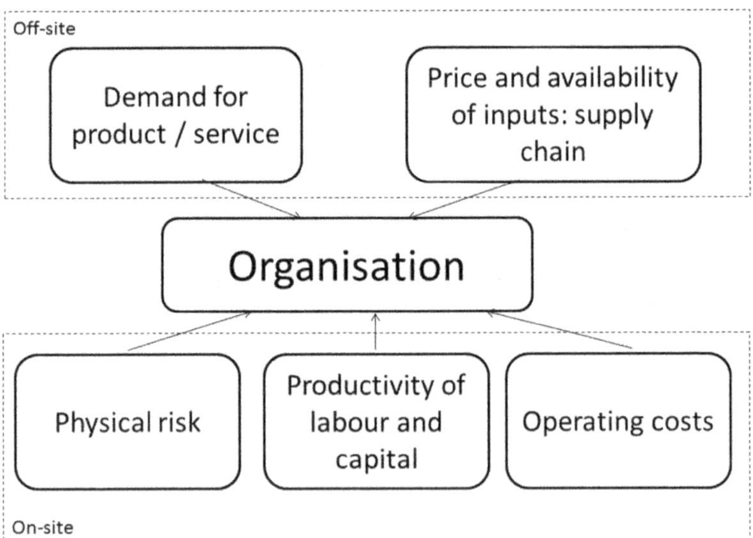

Figure 3.8 The impacts of climate change on an organisation

The bottom of the figure shows three aspects of an organisation that could directly be affected by climate change 'on-site'. Physical risk represents potential changes to the likelihood of damage to the organisation's sites, and to the likelihood of disruption to its activities (for example, because employees cannot get to work, customers cannot get to the site or distribution routes are closed). Operating costs include the costs of buying energy, water and insurance. Energy costs may be altered by changes in the demand for energy within the organisation, and insurance costs may change because the organisation's insurers judge that risk has changed. External factors may also affect energy and water costs; they may be affected by changes due to climate change in the costs incurred by energy and water suppliers in providing their services. The productivity of labour could be affected by changes in temperature or extreme weather conditions; these could be either negative or positive. The productivity of some types of equipment can also potentially be affected by weather conditions (increased need for cooling may lower efficiency, for example, and some chemical and biological processes are temperature-sensitive).

The demand for an organisation's products or services and the price and availability of raw materials can also be affected by climate change, and in these cases the relevant changes may be well away from any of the organisation's physical sites. Many products and services are weather sensitive (and there is correspondingly a considerable consulting industry providing weather data to organisations). At the level of the consumer, for example, weather affects demands for certain types of food, clothing and recreation, and it also affects demand for health services. Much of this sensitivity relates to periods of 'unusual' weather, and climate change is most obviously relevant if it changes the frequency of unusual weather events. At the business-to-business level, demand for heating and cooling equipment, repair and renovation services and irrigation equipment are sensitive to both weather and climate. For some products and services, climate change becomes relevant if consumers believe that climate change would mean that they would need those products and services more in the future. Examples are demand for air conditioning and irrigation equipment.

Perhaps the major implication of climate change for an organisation, however, is on its supply chain. The 2011 floods in Thailand had a well-documented impact on the global electronics industry, as production of vital components was severely disrupted. Seventy per cent of companies surveyed by the Carbon Disclosure Project (CDP) in 2013 identified a current or future business continuity risk related to climate change.[13] Climate change has the potential to affect the supply chain through impacting upon the availability of raw materials, their cost of production and transport costs. Sensitivity to supply chain disruption depends partly on the magnitude of the climate change impact on the cost and availability of raw materials, and partly on the concentration of sources of supply.[14] Agricultural commodities are most obviously directly affected by climate change, but as the Thailand floods example illustrates, even components that appear to be independent of weather can be affected if their production sites are exposed to disruption. Of course, there are also potential opportunities from climate change. The productivity of some crops is likely to increase, and it may become feasible to grow crops in areas that are currently

13 Carbon Disclosure Project (2013), *CDP Supply Chain Report 2012-13*. www.cdproject.net.

14 Gledhill, R., Hamza-Goodacre, D. and Low, L.P. (2013), Business-not-as-usual: tackling the impact of climate change on supply chain risk. *Resilience: a journal of strategy and risk*. www.pwc.com/resilience.

infeasible; costs could therefore come down, and availability increase. In the longer term – from the 2030s – it is possible that reductions in the extent of sea ice cover in the Arctic Ocean could shorten travel times (and therefore costs) between Europe and Asia, but this does depend on precisely how rapidly the sea ice shrinks and how hazardous the journey remains.

REDUCING IMPACTS THROUGH ADAPTATION

The impacts of climate change shown in Table 3.1 assume that nothing is done to seek to reduce those consequences. In practice, of course, this is unlikely to happen. The future impacts of climate change will be reduced by adaptation, defined by the IPCC as 'the process of adjustment to actual or expected climate and its effects'.[15]

The estimated impacts of climate change on flood losses in Europe shown in Table 3.1 assume no change in the provision of flood defences. New defences will be built as climate changes so the actual change in losses will be smaller than that given in the Table, although there will of course be costs involved in providing improved protection (in practice, the level of additional protection will likely be strongly influenced by the relative costs and benefits of improvements). The increases in cooling energy demands shown in Table 3.1 will also likely be alleviated by improved building design and increases in the efficiency of energy use, and changes in building design may also offset a proportion

15 IPCC (2014), Summary for Policymakers. In: *Climate Change 2014: Impacts, Adaptation and Vulnerability. Contribution of Working Group II to the Fifth Assessment Report of the Intergovernmental Panel on Climate Change*. Field, C., et al., eds. Cambridge University Press.

of the increased heat-related mortality. Reductions in crop yields may be offset by the use of crop varieties that are more resistant to high temperatures or drought.

Organisations and societies are used to attempting to cope with periods of unusual weather and climatic variability, for example by building flood defences or developing contingency plans, and this is typically based on past experience. Adaptation essentially involves changing these plans and approaches in anticipation of changes to climate and its variability. It is possible that measures put in place to cope with current climatic variability are sufficient to cope with what may happen under a changed climate – but that is not always necessarily the case. A flood defence structure may protect against floods with a return period of up to 100 years, and this would continue to provide some protection even if floods become more frequent. However, it would not provide the same *standard* of protection, and the chance of the defences being overtopped could increase: this would affect the risk to which properties behind the defences are exposed. The similarities and differences between responding to climatic variability and adapting to climate change are extremely important, and we will return to them in Chapter 6.

Some organisations that are particularly attuned to weather and its variability may adapt 'spontaneously' to climate change without explicitly deciding to do so. Many farmers decide (at least partly) what crops to grow the following year based on recent experience, and if that experience includes some climate change trend (for example for drier summers) then that will influence planting decisions. An organisation that can respond very quickly to altered conditions and has short pay-back times for investments may also 'spontaneously' adapt to climate change. However, most organisations would need to consider explicitly whether or not to adapt to climate change: their strategies, operations and investments make assumptions about risks and opportunities over several years.

In practice there are many barriers to adaptation. These are summarised in Figure 3.9. Without some awareness by key players in an organisation of the threats and opportunities posed by climate change, there is unlikely to be much attempt to adapt to it. Key players include senior executives, board members and external stakeholders such as shareholders; those in a position to make decisions need to be aware of the implications of climate change. Several factors affect awareness. These include understanding and acceptance of the evidence for climate change, experience and interpretation of

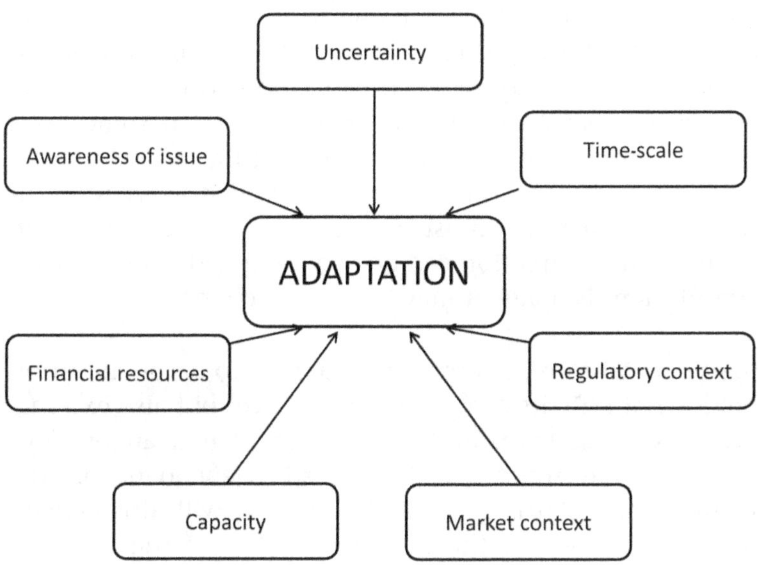

Figure 3.9 Barriers to adaptation

recent extreme events, and confidence in methods to estimate possible future changes. A perception that climate is changing will affect how an organisation reacts to an extreme event: the response is likely to be different if the event is seen as a 'one-off' or as indicative of an ongoing trend. The awareness of potential future threats and opportunities also influences the extent to which an organisation is prepared to anticipate future climate change and adapt proactively.

Time-scale is important, and can be a barrier to effective adaptation in two contrasting ways. First, the scales over which climate change – as opposed to climatic variability – may affect an organisation may be considerably longer than time-scales over which the organisation typically plans and believes itself able to change. These planning time-scales obviously vary from organisation to organisation. If it is believed that climate change will have an impact too far into the future, then the organisation is unlikely to adapt; this could expose an organisation to increased risk. Second, in complete contrast, it is possible in some cases that the impacts of climate change could be seen *before* an organisation is able to respond. This is most likely to occur in sectors – such as infrastructure – where lead times from inception to completion can be very long and where, once a commitment is made, it may be hard to change.

The financial resources that are available to an organisation include not only its own internal resources but also external sources such as loans and investments. The availability of these external sources will be partly dependent on the organisation's status and reputation, but will also depend on the awareness of the external funders of the need for investment and the credibility of the organisation's case for investment.

The capacity of an organisation to adapt depends on both its organisational culture and the skills and knowledge of its staff. Relevant aspects of organisational culture include attitudes to change, the presence or absence of rigid prescribed ways of doing things, and concern about reputation; the nature of the organisation's 'licence to operate' amongst its stakeholders may also be important.

Many aspects of the way an organisation operates and adapts to climate change may be strongly constrained by market and regulatory context. The market may constrain adaptation, by making it difficult for an organisation to do anything that involves increasing costs. On the other hand, the market may encourage adaptation through competitive behaviour – if an organisation that is seen to adapt is perceived to have a competitive edge. Similarly, the regulatory framework within which an organisation operates can stifle or encourage adaptation. If the framework discourages innovation or forces particular ways of operating, then adaptation is likely to be more difficult. In contrast, a regulatory framework which requires organisations to consider risks, and more particularly the effect of climate change, has the potential to encourage adaptation (we will return to examples of such frameworks in Chapter 4).

However, even in a highly supportive environment – strong awareness, available resources and capacity, and an enabling regulatory and market context – uncertainty poses a major barrier to adaptation. There are two major components to this uncertainty: uncertainty in what to adapt to (i.e. how will climate change affect the organisation), and uncertainty in what to do. We have already seen in Chapter 2 and in this chapter how it is difficult to make precise quantitative projections of how climate change will impact upon a specific sector in a specific place. Sometimes we can estimate the broad direction of change – sea level will go up and temperatures will rise – but in other cases even the direction of change (more flooding/less flooding) can be uncertain. This obviously makes it difficult to plan an adaptation strategy, and the uncertainty may be believed to be so large that there is little point in trying to do anything now: why not wait until the uncertainty is reduced? In practice, it is unlikely that uncertainty in projected changes will ever be eliminated, so 'wait and see' is not necessarily an appropriate option. Organisations also may not be aware of the full range of adaptations that can be made, or indeed how appropriate different responses are, given the uncertainty in how risks may change. Dealing with uncertainty in how climate will change and impact upon an organisation is fundamental to managing climate change risk, and we will therefore return to this in more detail in Chapters 5 and 6.

Organisations and individuals will reduce the impacts of climate change through adaptation, although the extent and effectiveness of adaptation is difficult to predict and there are many barriers to overcome. There is also the risk of 'maladaptation', defined as a response which rather than reducing vulnerability increases it instead. This can take several forms. For example, an organisation may seek to adapt to increased flood risk by building a higher flood wall around some critical infrastructure, and then – because the site is now 'safe' – increasing investment on the site. If it turns out that the new flood wall is not large enough then these more valuable assets are exposed to a greater likelihood of flooding and the risk to the organisation is increased. Another example would be where an organisation decides to stop worrying about a particular risk because it is less likely to happen – only to be caught out when an event strikes. With higher temperatures, for example, we might expect to see less snowfall and therefore fewer disruptive events, so we would not bother to maintain snowploughs or plans. However, we could still get disruptive snowfalls – albeit less frequently – in a warmer world, and if an organisation is less prepared than currently, it will be more severely affected.

CONCLUSION

This chapter has shown the potential impacts of climate change on many sectors and aspects of society, and how climate change may impact both directly and indirectly on an organisation. These impacts may pose threats, or may provide new opportunities. However, projections of what these impacts are uncertain, for several reasons:

- We do not know exactly how future emissions will change (although over the next three or four decades this is not too important);
- We do not know how regional and local climate will change in response to increasing emissions; and
- We do not know precisely how changes in climate will impact upon sectors and organisations.

This uncertainty is important, and in fact it determines how we should assess and respond to climate change risks. Meanwhile, of course, other things are changing too. Population is generally increasing (and, in some countries, ageing), economic growth (or lack of it) alters production and consumption, and innovation brings new products to the market. Some of these trends also affect the environment, through land use change, or through increasing demands for environmental protection. Climate change has to be seen in the context of these other drivers of change.

Governments, organisations and individuals are not likely to sit back and wait to be impacted by climate change. However, these adaptations to climate change will come at a cost and, as we have seen, there are some significant potential barriers to adaptation.

Governments and regulators provide a framework within which organisations respond to the challenges of climate change. Since the early 1990s there have been global initiatives to seek to reduce emissions of greenhouse gases, and over time there has been increasing interest at international, national and local levels in supporting adaptation to climate change. The next chapter summarises how these policy frameworks have evolved, and what they imply for organisations seeking to manage their climate change risks.

4 The Politics of Climate Change: The Regulatory Context

INTRODUCTION

International organisations, national and regional governments, and regulatory agencies have all over the last few years developed policies and procedures to address climate change. Whilst most of these actions have focused on ways of reducing future emissions of greenhouse gases ('mitigation'), many are also concerned with supporting and encouraging adaptation.

This chapter provides an overview of the political and regulatory framework within which organisations will respond to climate change, looking first at international climate policy, then at national and regional legislation, and finally at initiatives developed by regulators and other bodies. This context is important for several reasons. Policy and regulations – and particularly the debates around them – raise awareness of climate change as an issue within an organisation and amongst its stakeholders. Policy and regulations may make organisations do things that they would not otherwise have done, and may either facilitate or constrain actions that organisations take. They allow benchmarking of actions between organisations, for example through reporting requirements. They also provide a context for organisations to enhance their reputations amongst stakeholders – by being seen to 'do the right thing' – but also provide a source of reputational risk if organisations are perceived to be not living up to expectations. Finally, the policy and regulatory framework is evolving, and uncertainty in how policy and regulations will change in the future represents an additional layer of risk to an organisation – regulatory risk. This regulatory risk may strongly influence how an organisation deals with the risks of climate change outlined in Chapters 2 and 3. The focus is on policy and legislation relevant to climate risks, rather than mitigation in itself, so the chapter concentrates on issues such as overall climate targets and policies to support adaptation.

INTERNATIONAL CLIMATE POLICY: THE KYOTO PROTOCOL AND BEYOND

Most public debate and discussion around responses to climate change is focused around attempts to get international agreements to reduce the future emissions of greenhouse gases. The Kyoto Protocol, adopted in December 1997, sets targets for emissions reductions by developed countries, and a successor to the Kyoto Protocol is due to be agreed in Paris in December 2015. This international activity is built around the United Nations Framework Convention on Climate Change

(UNFCCC[1]), but – as will be outlined below – the UNFCCC covers much more than simply reductions in future emissions.

The next section summarises the UNFCCC process and key milestones, concentrating on mitigation, and the following section looks at the challenges to reaching an international agreement on emissions reduction. Then we look at how effective emissions reductions would be in reducing the increase in global mean temperature, before outlining the other key parts of the UNFCCC process focusing on adaptation and compensation for loss and damage.

THE UNFCCC PROCESS

The first report of the Intergovernmental Panel on Climate Change (IPCC) was presented to the Second World Climate Conference sponsored by the UN's World Meteorological Organisation (WMO) in Geneva in late October and early November 1990. This presented the first consensus assessment that human activities had led to increasing concentrations of greenhouse gases, and that continued emissions would lead to further increases in global mean temperature. As early as December 1990, negotiations started at the UN General Assembly to develop an international agreement to address climate change, and the draft of the UN Framework Convention on Climate Change was adopted on 9 May 1992 at New York. The UNFCCC was opened for signature at the UN Conference on Environment and Development (the 'Earth Summit') in Rio de Janeiro in June 1992, and came into force in March 1994 after receiving the necessary 50 national ratifications. It has now been ratified by all United Nations member states and the European Union.

1 Usually pronounced 'U N F triple C'.

The UNFCCC[2] is inevitably a compromise between different national interests. The objective of the UNFCCC is set out in Article 2:

> *The ultimate objective of this Convention ... is to achieve ... stabilization of greenhouse gas concentrations in the atmosphere at a level that would prevent dangerous anthropogenic interference with the climate system. Such a level should be achieved within a time frame sufficient to allow ecosystems to adapt naturally to climate change, to ensure that food production is not threatened and to enable economic development to proceed in a sustainable manner.*

This objective does not define 'dangerous anthropogenic interference', and of course the second sentence is open to many interpretations. Article 3 sets out the principles underpinning the UNFCCC. Article 3.1 states:

> *The Parties should protect the climate system for the benefit of present and future generations of humankind, on the basis of equity and in accordance with their common but differentiated responsibilities and respective capabilities. Accordingly, the developed country Parties should take the lead in combating climate change and the adverse effects thereof.*

Along with this recognition of different responsibilities,[3] Article 3.3 states that:

> *The Parties should take precautionary measures to anticipate, prevent or minimize the causes of climate*

2 www.unfccc.org.

3 The UNFCCC identifies one set of countries listed in 'Annex I' as having prime responsibility for reducing future emissions: these are the industrialised countries, but excluding South Korea, Brazil and other rapidly industrialising countries.

> *change and mitigate its adverse effects. Where there are threats of serious or irreversible damage, lack of full scientific certainty should not be used as a reason for postponing such measures, taking into account that policies and measures to deal with climate change should be cost-effective so as to ensure global benefits at the lowest possible cost.*

These Articles are quoted at length because they define not only the objectives of international policy but also the different responsibilities of different countries and the treatment of uncertainty; these are central to understanding how international policy has developed since the launch of the UNFCCC in 1992.

As an aside, it is worth noting that the UNFCCC is not in fact the first international climate policy. The Montreal Protocol was signed in 1987, with the aim of eliminating the production and use of chemicals that deplete the ozone layer; the observed reduction in the stratospheric[4] ozone layer was believed to allow increased penetration of UVB radiation, associated with increased risk of skin cancer. Some of these chemicals are also greenhouse gases, and these are not covered in the UNFCCC because they are already covered under the Montreal Protocol.[5]

4 The stratosphere is the second major layer in the earth's atmosphere, and starts at around 10-13km above the surface (it is lower at the poles).

5 Apart from the fact that some chemicals both deplete stratospheric ozone and act as greenhouse gases, there are other links between stratospheric ozone depletion and climate change caused by increasing greenhouse gas concentrations. One effect of increasing greenhouse gas concentrations is a cooling of the stratosphere, which leads to stratospheric ozone depletion; on the other hand, ozone depletion cools the troposphere below because the cooler stratosphere emits less radiation downwards. The 'greenhouse' warming effects of stratospheric ozone-depleting chemicals offset their cooling effects from the ozone depletion.

Table 4.1 Notable COPs

	When?	**Where?**	**Outcome**	**What does it do?**
COP3	1997	Kyoto	Kyoto Protocol	Sets first emission reduction targets
COP12	2006	Nairobi	Nairobi Work Programme	Sets up a process to support understanding of impacts, vulnerability and adaptation
COP13	2007	Bali	Bali Road Map	Sets new negotiating process designed to tackle climate change, including the Bali Action Plan
COP15	2009	Copenhagen	Copenhagen Accord	Defines aspirations of governments and encourages them to pledge commitments
COP16	2010	Cancun	Cancun Agreements	Establish goals and schedule for reducing emissions, and establishes review of 2°C target. Includes Cancun Adaptation Framework
COP17	2011	Durban	Durban Outcomes	Governments committed again to attempt to reach agreement
COP18	2012	Doha	Doha Gateway	Some commitments for reductions in emissions over the period 2013-2020.
COP21	2015	Paris	??	Target meeting for new agreement

The UNFCCC defined some vague targets for the reduction of emissions (in Article 4.2b), but more importantly set up an institutional process managed by a secretariat based in Bonn, Germany. This process includes a periodic Conference of the Parties (COP) to the UNFCCC and two bodies providing advice to the UNFCCC. The Subsidiary Body on Scientific and Technical Advice (SBSTA) does what its name suggests, and this is what the IPCC now reports to. The Subsidiary Body on Implementation (SBI) provides technical advice on the implementation of the UNFCCC.

However, whilst the technical details are discussed at SBSTA and SBI meetings, the key meetings of the UNFCCC are the COPs, which have since 1997 been held annually. Table 4.1 summarises some of the outcomes from particularly significant COPs.

THE KYOTO PROTOCOL AND BEYOND

The Kyoto Protocol was agreed at COP3 in 1997, and is the cornerstone of international climate policy. In essence, the Kyoto Protocol sets emissions reduction targets for developed countries (known as Annex B countries[6]). These targets were intended to lead to an overall reduction in emissions over the period 2008-2012 of at least 5% relative to 1990 emissions. The targets are legally binding, and apply to what is known as a basket of four greenhouse gases (CO_2, CH_4, N_2O and SF_6)[7] and two groups of flourinated gases. Table 4.2a summarises the obligations of Annex B countries. The European Union is a signatory, and has the scope to produce different targets for each member state through what is known as an 'effort-sharing' agreement, subject to the overall reduction target of 8%. Table

6 All UNFCCC Annex I countries except Belarus and Turkey.

7 Carbon dioxide, methane, nitrous oxide and sulphur hexaflouride.

Table 4.2a Emissions limitation and reduction targets for 2008-2012 for Annex B countries under the Kyoto Protocol

Country	Target (% limitation or reduction relative to base year)
Austria, Belgium, Bulgaria, Czech Republic, Denmark, Estonia, Finland, France, Germany, Greece, Ireland, Italy, Latvia, Liechtenstein, Lithuania, Luxembourg, Monaco, Netherlands, Portugal, Romania, Slovakia, Slovenia, Spain, Sweden, Switzerland, UK, plus the European Union (as of 1997)	-8%
USA	-7%
Canada, Hungary, Japan, Poland	-6%
Croatia	-5%
New Zealand, Russia, Ukraine	0
Norway	+1%
Australia	+8%
Iceland	+10%

Table 4.2b Emissions limitation and reduction targets for 2008-2012 for the 15 countries of the EU in 1997

Country	Target	Country	Target
Austria	-13%	Italy	-6.5%
Belgium	-7.5%	Luxembourg	-28%
Denmark	-21%	Netherlands	-6%
Finland	0%	Portugal	+27%
France	0%	Spain	+15%
Germany	-21%	Sweden	+4%
Greece	+25%	UK	-6.5%
Ireland	+13%		

4.2b shows the individual targets for the 15 countries that were members of the EU in 1997. The emissions targets clearly represent the end result of horse-trading between countries, and were not designed with a specific climate target in mind. The base year for reductions in the key greenhouse gases CO_2, CH_4 and N_2O is 1990 (although some countries in eastern Europe could specify other base years), but for the other gases the base year is 1995. The 'commitment period' is 2008-2012, meaning that emissions are averaged over five years in order to smooth out any potentially anomalous years (for example ones with unusual weather and hence energy demands, or with unusual economic conditions). Individual countries had to sign and ratify the Kyoto Protocol, and it formally entered into force in February 2005 after countries responsible for at least 55% of total CO_2 emissions in 1990 had ratified. By 2013, 191 countries plus the EU had ratified the Kyoto Protocol. The United States never ratified, and Canada withdrew in 2012.

The Kyoto Protocol allowed the countries with targets to meet these targets in five ways. The first is simply by reducing domestic emissions of greenhouse gases from the combustion of fossil fuels and other activities. The second is through increasing the sinks of carbon by planting trees and managing agricultural land. This is known in the language of the UNFCCC as LULUCF (land use, land use change and forestry), and countries can count activities since 1990 against their emissions targets. The third and fourth mechanisms allow developed countries to gain emissions reduction credits by supporting approved emissions reduction measures in other countries. The Clean Development Mechanism (CDM) is specifically designed to support activities in developing countries, whilst Joint Implementation (JI) is designed to support deals between industrialised countries (including those that do not have specific targets). To make things slightly more complicated, both CDM and JI projects can include measures to reduce emissions from fossil fuel burning and through LULUCF. The final mechanism countries can use to achieve targets under the Kyoto Protocol is through the trading of emissions permits. Emissions trading involves the setting of a total cap on emissions and the sale of permits to emit to (and between) emitters. It is designed to achieve target emissions reductions in the most economically effective way. The European Union Emissions Trading System (ETS) is currently the largest in operation. The Kyoto Protocol is legally binding in the sense that countries which are deemed not to be compliant are exposed to some form of sanction. In principle these could be a requirement to 'make up the difference' in a later commitment period, or exclusion from some of the Kyoto Protocol mechanisms.

Clearly, all these mechanisms need specific procedures and agreed rules, so much of the discussion in COPs after 1997 has focused around implementation. Monitoring, reporting and verification (known as MRV) is particularly important, especially around the area of LULUCF.

So how effective has the Kyoto Protocol been? Verified data are only available to 2011,[8] but overall Annex B countries have reduced their total emissions by 2008-2011 compared with 1990 by around 14%, or 9% if LULUCF is not included (Figure 4.1a) – considerably more than the 5% target. However, whilst the EU has exceeded its Kyoto Protocol target, most of the decrease occurred in Russia and other eastern European countries where large reductions in emissions occurred not because of deliberate climate policies but because of major economic restructuring. The USA missed its target (although

8 On the UNFCCC website: www.unfccc.org.

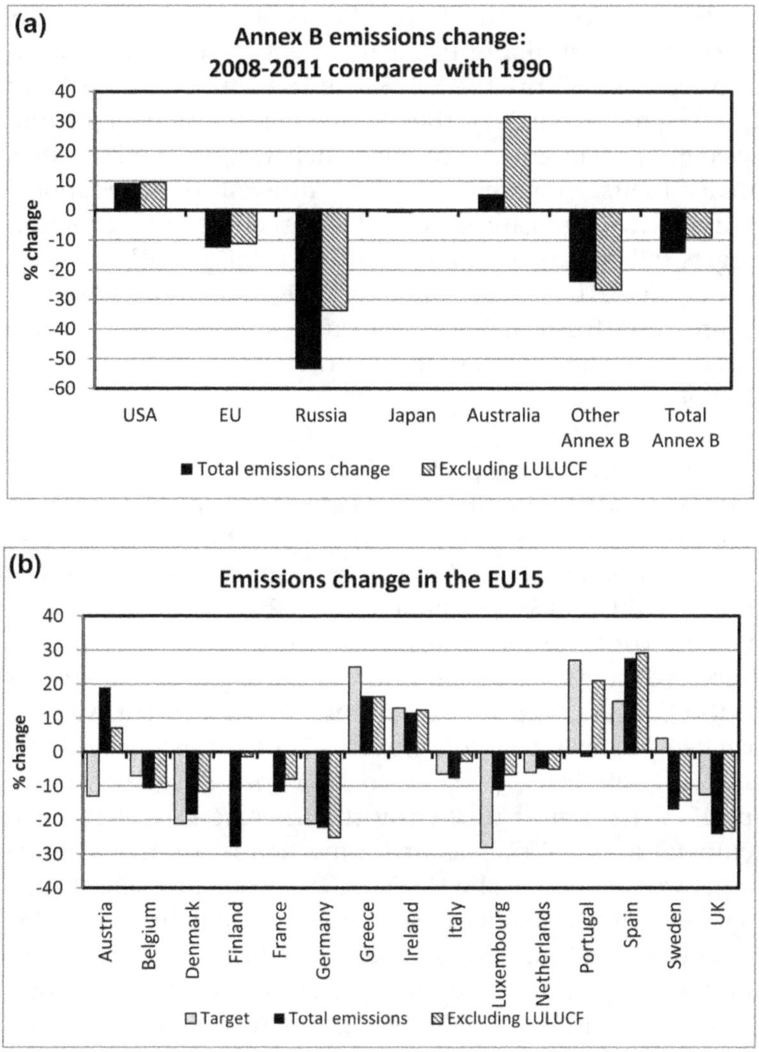

Figure 4.1 **Emissions from countries with targets under the Kyoto Protocol: (a) all countries; and (b) countries within the European Union. Data from www.unfccc.org**

it does not follow the Kyoto Protocol), Japan missed its target, and Australia only hit its target if LULUCF is taken into account. Within the EU15 (the 15 members of the EU in 1997), some countries – the UK, Germany, France, Belgium, Sweden and Finland – exceeded their targets, whilst others missed theirs. However, whilst the Annex B countries appear to have in aggregate exceeded their emissions reduction targets (albeit largely inadvertently), total global emissions of CO_2 in 2010 had increased by 31% relative to 1990 (see Figure 2.3)[9] (fossil fuel emissions increased by 42%, but this was offset by a reduction of 11% in emissions from land use change).

The first commitment period of the Kyoto Protocol ended in December 2012. The process to extend or replace the Kyoto Protocol was started at COP13 in Bali in 2007 and specified in the Bali Road Map. The original aim was to agree a new protocol at COP15 in Copenhagen in 2009, but this did not happen. For a number of reasons – including the developing global economic recession and domestic political concerns in the USA and China – a binding new agreement was not reached. Instead, the non-binding Copenhagen Accord was adopted, which has two key components. First, the Accord specifies the aim of restricting the increase in global mean temperature to 2°C, the first time a quantitative climate target was defined in the UNFCCC process. The Copenhagen Accord does not explicitly state a reference date for the 2°C rise, but this is interpreted to be 'pre-industrial'; global mean surface temperature over the period 1986-2005 is around 0.6°C higher than pre-industrial, implying that the 2°C target

9 From data in IPCC (2014), Summary for Policymakers. In: *Climate Change 2014: Mitigation of Climate Change. Contribution of Working Group III to the Fifth Assessment Report of the Intergovernmental Panel on Climate Change*. Edenhofer, O., et al., eds. Cambridge University Press.

is equivalent to a rise in global mean surface temperature of around 1.4°C above current levels. The Accord also does not state when the target should be reached: is it never, or by the end of the 21st century? If the latter, does it matter if temperatures rise above 2°C before falling back below the target? These appear to be subtle distinctions, but they make a difference when determining whether a given emissions pathway is likely to achieve the 2°C target.

The second key part of the Copenhagen Accord is that governments agreed to make pledges of future targets for the reduction in greenhouse gas emissions by 2020 (Table 4.3). Governments have made a wide range of pledges, many conditional on the adoption of a new binding agreement. The EU, for example, pledged a reduction in emissions of 20% by 2020 (relative to 1990), and a reduction of 30% if other countries joined in. The US pledged a reduction of 17%, but relative to 2005 rather than 1990. For the first time, countries outside Annex B also made quantitative pledges. Most of these – including those of Brazil, Mexico and South Africa – are to reduce emissions in 2020 relative to what they would be in the absence of any policy (requiring of course an estimate of what those emissions would be). Some – including those of China and India – are in the form of pledges to reduce the carbon emissions per unit of GDP.

COP16 in Cancun, Mexico, in 2010 approved the Cancun Agreements, which apart from restating the need to produce a new binding agreement also proposed an 'aspirational' temperature change target of 1.5°C (or approximately 0.9°C above current levels) and initiated a review of the long-term temperature target (due to report in 2015). At COP18 in Doha in 2012 a number of countries made emissions reduction commitments for the second commitment period

Table 4.3 Quantitative emissions reduction pledges under the 2009 Copenhagen Accord

	Emissions reduction in 2020	**Base year**
Annex I		
Australia	Unconditional offer of 5% cut by 2020, 15% with a partial global deal, and 25% with an 'ambitious' global deal (450ppm)	2000
Belarus	5-10% reduction, conditional on access to funding and mechanisms	1990
Canada	17%, to be aligned with the USA	2005
EU and member states	Unconditional 20% reduction, 30% reduction with global deal	1990
Iceland	30% reduction, as part of a global deal	1990
Japan	25% reduction, as part of a global deal	1990
Kazakhstan	15%	1992
Liechtenstein	Unconditional 20% reduction, 30% reduction with global deal	1990
Monaco	30%	1990
New Zealand	10-20% reduction, as part of a global deal	1990
Norway	30-40% reduction, as part of a global deal	1990
Russian Federation	15-25% reduction, as part of a global deal	1990
Switzerland	Unconditional 20% reduction, 30% reduction with global deal	1990
Ukraine	20%, conditional on access to funding and mechanisms	1990
USA	'in the range of 17%', depending on legislation	2005

Non-Annex I		
Antigua and Barbuda	25% reduction	1990
Brazil	36-39% reduction, relative to 'business-as-usual' emissions	2020
Chile	20% reduction, relative to 'business-as-usual' emissions	2020
China	40-45% reduction in carbon dioxide emissions per unit of GDP	2005
Costa Rica	'significantly deviate' from business-as-usual emissions in a 'long-term economy-wide transformational effort to enable carbon-neutrality'	
India	20-25% reduction in carbon dioxide emissions per unit of GDP	2005
Indonesia	26% reduction	?
Israel	20% reduction, relative to 'business-as-usual' emissions	2020
Kyrgyz Republic	20% reduction, relative to 'business-as-usual' emissions	2020
Maldives	Achieve carbon neutrality by 2020	
Marshall Islands	40% reduction	2009
Mexico	30% reduction, relative to 'business-as-usual' emissions	2020
Moldova	25% reduction, as part of a global deal	1990
Papua New Guinea	50% reduction before 2030, while becoming carbon neutral before 2050	2030
Singapore	16% reduction, relative to 'business-as-usual' emissions	2020
South Africa	34% reduction, relative to 'business-as-usual' emissions	2020
South Korea	30% reduction, relative to 'business-as-usual' emissions	2020

2013-2020 (Table 4.4), although these are not yet binding. The EU committed to a 20% reduction in emissions relative to 1990 – again with different responsibilities across member states – and Belarus, Kazakhstan, Norway, Switzerland and the Ukraine also committed to reductions. However, three of the countries with original commitments under the Kyoto Protocol – Japan, New Zealand and Russia – declined to make quantitative commitments for 2013-2020, and of course neither the USA nor Canada now follow the Kyoto Protocol.

Table 4.4 Quantified committed reductions in emissions by Annex B countries under the second commitment period of the Kyoto Protocol, as agreed at COP18 in Doha

	Quantified emissions reductions in second commitment period 2013-2020	Base year
Australia	2%	2000
Belarus	12%	1990
EU and member states	20%	1990
Iceland	20%	1990
Kazakhstan	5%	1990
Liechtenstein	6%	1990
Monaco	22%	1990
New Zealand	To be defined	
Norway	6%	1990
Switzerland	15.8%	1990
Ukraine	24%	1990

Note: Japan and the Russian Federation declined to make quantified commitments, and the USA and Canada are outside the Kyoto Protocol. The EU includes all 27 member states, in contrast to the 15 countries at the time of the agreement of the Kyoto Protocol

Currently, the aim of the UNFCCC process is to reach a new agreement at COP21 in Paris in 2015.

CHALLENGES TO THE UNFCCC PROCESS: WHERE ARE THE PROBLEMS?

Although the Kyoto Protocol was agreed in 1997, it has clearly been difficult to follow this up with new arrangements. There are several reasons for this. The UNFCCC process has become complex and cumbersome, and involves negotiations between many groups of countries with different perspectives and agendas. Many of these groups are themselves divided over specific issues. The UNFCCC covers many issues beyond the reduction of future emissions: these are discussed later in this chapter, and in practice occupy much negotiating time. There are a number of significant technical issues, such as how to account for the effects of land use change on a country's carbon emissions and how to seek to limit or reduce emissions used in international air and marine transport.

There are, however, two key issues of principle that are currently hindering the development of a new binding global policy to reduce emissions. The first relates to the target for the maximum increase in global mean surface temperature, and the second relates to the allocation of restrictions on future emissions.

As outlined above, neither the UNFCCC itself nor the Kyoto Protocol specified a target for the maximum 'allowable' amount of climate change, but the Copenhagen Accord of 2009 introduced a target of limiting the increase in global mean surface temperature to less than 2°C above preindustrial levels (about 1.4°C above current levels). The Cancun Agreement in 2010 reaffirmed this commitment, but also initiated a review of what is known as the long-term global goal, with specific reference to a potential lower goal of 1.5°C (0.9°C above current levels). The temperature goal affects the emissions reductions that are necessary.

The Kyoto Protocol just imposed restrictions on future emissions from the 'Annex B' countries listed in Table 4.2a. These do not include many rapidly industrialising countries. In 1990 the Annex B countries accounted for approximately 65% of global emissions of CO_2 from fossil fuel combustion,[10] but by 2009 only accounted for 43%. The proportion is projected to fall further. Whilst the UNFCCC and Kyoto Protocols recognise 'differentiated responsibilities' between different countries, most of the debate about a future binding agreement revolves around what this means in practice. Should the major emitters outside Annex B (including China, India, Brazil, Mexico, South Korea, Indonesia and South Africa) be subject to limitations on future emissions? Should future limits aim for some form of per capita convergence, or should historical emissions be taken into account (this is known as the 'Brazilian proposal')?[11] The resolution of these major questions has to be political.

WHAT EFFECT WOULD REDUCING EMISSIONS HAVE ON CLIMATE CHANGE?

The climate system responds slowly to the increasing concentration of greenhouse gases in the atmosphere, so the temperature change by a particular point is strongly related to the accumulated emissions up to that point. This is illustrated in Figure 4.2, which shows the relationship

10 Data from http://cdiac.ornl.gov/trends/emis/annex.html.

11 It is, however, difficult to estimate historical contributions because data are poor and results are sensitive to assumptions.

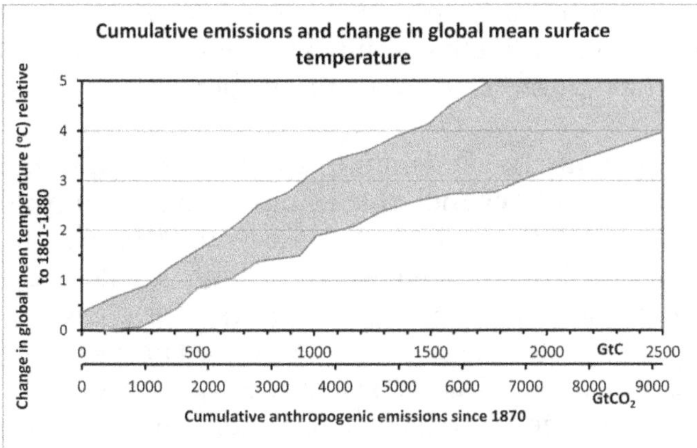

Figure 4.2 The relationship between cumulative emissions of CO_2 and change in global mean temperature. Note that cumulative emissions do not include other greenhouse gases. Redrawn from the IPCC AR5 Working Group I Summary for Policymakers

between cumulative emissions of CO_2 from 1870 against the rise in global mean surface temperature (relative to pre-industrial levels). The graph shows that if we emit a total of 1000 Gigatonnes (a trillion tonnes) of carbon (equivalent to 3667 Gigatonnes of CO_2) through the combustion of fossil fuels and land clearance into the atmosphere, then the global average surface temperature will rise by somewhere between 1.5°C and 2.8°C. Looked at the other way, if we want to limit the rise in temperature to 2°C, then we must emit less than 1210 GtC to have a 50% chance of achieving our target, or less than 1000 GtC to have a greater than 66% chance of keeping the temperature rise below 2°C.[12] We have already emitted approximately 515 GtC (with an uncertainty

12 IPCC (2013), Summary for Policymakers. In: *Climate Change 2013: The Physical Science Basis. Contribution of Working Group I to the Fifth Assessment Report of the Intergovernmental Panel on Climate Change*. Stocker, T., et al., eds. Cambridge University Press.

range of 445 to 585) since 1870. If we factor in the other greenhouse gases (methane, nitrous oxide and so on), then our 'carbon budget' is reduced still further.

Limiting cumulative emissions to around a trillion tonnes to have a 'reasonable' chance of keeping the rise in temperature below the 2°C target is consistent with keeping annual emissions in 2020 to somewhere around 42-46 $GtCO_{2eq}$[13] (taking into account all greenhouse gases). The pledges made by countries under the Copenhagen Accord imply emissions in 2020 of between 47 and 53 $GtCO_{2eq}$,[14] which would not be enough to hit the 2°C target. UNEP terms this difference (of approximately 8-12 $GtCO_{2eq}$/year in 2020) the 'emissions gap'.[15]

In order to achieve the 2°C target it is necessary either to make greater reductions sooner, or make much more rapid reductions after 2020. It is also likely that achieving the target will require 'negative emissions', where human activities absorb more carbon than they emit. Such negative emissions could in principle be achieved by reforestation and carbon capture and storage, but there are very many technical and political problems with both.[16]

This is not the place to review in detail whether it is technically, economically or politically feasible to achieve the substantial

---

13 den Elzen, M.G., et al. (2011), The emissions gap between the Copenhagen pledges and the 2°C climate goal: options for closing and risks that could widen the gap. *Global Environmental Change* 21: 733-743.

14 den Elzen, M.G., et al. (2011), The emissions gap between the Copenhagen pledges and the 2°C climate goal: options for closing and risks that could widen the gap. *Global Environmental Change* 21: 733-743; Rogelj, J., et al. (2010), Analysis of the Copenhagen Accord pledges and its global climatic impacts: a snapshot of dissonant ambitions. *Environmental Research Letters* 5.

15 UNEP (2013), *Emissions Gap Report 2013*. United Nations Environment Programme (UNEP), Nairobi.

16 UNEP (2013), *Emissions Gap Report 2013*. United Nations Environment Programme (UNEP), Nairobi.

reductions in emissions necessary to close the 'emissions gap' in order to achieve UNFCCC climate targets. It is sufficient to note that several studies have demonstrated that it is still feasible – technically – to make significant reductions globally,[17] but that the window of opportunity is narrowing. For example, achieving the UN's goal of 'Sustainable Energy for All' would mean achieving the 2°C target.[18] This goal involves doubling the share of renewable energy in the global energy mix by 2030 and doubling the rate of improvement in energy efficiency, whilst ensuring universal access to modern energy services.

Because the climate system responds slowly to increasing greenhouse gas concentrations, it will take time for any reductions in emissions to have an effect on global temperature. This can be seen in Figure 2.5 where the changes in global mean temperature under the scenario with the greatest emissions are similar to at least 2030 to those under the scenario with the lowest emissions (and note that this lowest scenario assumes vary rapid reductions in emissions). Sea level responds even more slowly than temperature. This means that even if we were able to make significant reductions in emissions – for example to achieve the 2°C target – then there would be little clear reduction in the impacts of climate change for the next few decades. For example, if emissions were to follow a pathway to achieve the 2°C target, the numbers of people globally exposed to increased water scarcity would be reduced by only 5-8%, although the numbers exposed to increased flood risk would be reduced by 30-50%.[19] However,

17 UNEP (2013), *Emissions Gap Report 2013*. United Nations Environment Programme (UNEP), Nairobi.

18 Rogelj, J., et al. (2013), The UN's 'Sustainable Energy for All' initiative is compatible with a warming limit of 2°C. *Nature Climate Change* 3: 545-551.

19 Arnell, N.W., et al. (2013), A global assessment of the effects of climate policy on the impacts of climate change. *Nature Climate Change* 3: 512-519.

reducing emissions does delay significantly the impacts of climate change.

OTHER COMPONENTS OF THE UNFCCC PROCESS

Although the primary focus of the UNFCCC process has been on curbing the increase in emissions of greenhouse gases, the UNFCCC has many other strands.[20] Countries are required to report their actions to deal with climate change, with different requirements for Annex I (i.e. developed) and 'non-Annex-I' countries. Annex I countries have to submit National Communications every four years, and the sixth is due in early 2014. By the middle of 2013, just over 100 non-Annex I countries had submitted a second National Communication, and a small number had submitted more. These National Communications provide information on national conditions affecting exposure to, and response to, climate change, along with information on emissions, policy measures in place and (particularly for non-Annex I countries) exposure and vulnerabilities. Developing countries have also submitted reports describing Nationally Appropriate Mitigation Actions (NAMAs); these are mostly lists of generalised or specific actions to increase energy efficiency, enhance the use of renewable energy sources, and reduce deforestation. National Communications and NAMAs provide valuable information on a country's exposure to, and response to, climate change.

Whilst most of the mitigation effort has been directed towards reducing emissions from the burning of fossil fuel, a considerable amount of attention has been spent on measures to reduce emissions from land. The Kyoto Protocol explicitly includes LULUCF, but particularly since the COP16 in Cancun

20 See www.unfccc.org for more information.

there has been increasing interest in the management of forest. This is known as REDD-plus (Reducing Emissions from Deforestation and forest Degradation, plus conservation and enhancement of forest carbon stocks and the sustainable management of forests). Most of the discussion is around techniques for assessing and evaluating REDD-plus schemes, but in some countries these may be a major part of policies to reduce emissions – and have knock-on effects for other land uses.

Adaptation to climate change has risen gradually up the UNFCCC agenda. The Nairobi Work Programme established in 2006 set up a process to support understanding of adaptation in developing countries, and a programme to construct National Adaptation Programmes of Action (NAPAs) was initiated in 2001. By the middle of 2013, 50 developing countries had submitted NAPAs, which contain specific measures to address vulnerabilities to climate change. The Bangladesh NAPA, for example, contains 15 prioritised actions, which are a combination of awareness raising, capacity building and specific interventions (e.g. to reduce risk through afforestation along the coastal zone). The Angolan NAPA lists 29 specific actions, and the Ethiopian NAPA uses a set of criteria to whittle a long list of 37 projects down to a priority list of 11. Specific actions in NAPAs are eligible for funding through the Global Environmental Facility (GEF) and the Adaptation Fund.

The GEF was set up by the World Bank in 1991 as a means of supporting investment in environmental projects.[21] It is now an independent body and acts as a funder for the UNFCCC (and other international conventions). It provides grants and leverages funding, and is replenished largely by

21 http://www.thegef.org/gef/whatisgef.

contributions from individual countries. The Adaptation Fund was established under the UNFCCC in 2001, and focuses on funding specific adaptation measures, again in newly industrialising or developing countries. Much more ambitious is the Green Climate Fund, established following a pledge by developed countries at the Cancun COP in 2010 to create a fund worth $100 billion a year by 2020 to support mitigation and adaptation in the developing world. Discussions in subsequent COPs have focused on how actually to extract that level of funding from donors.

There is also a strong technology component to the UNFCCC, with a specific focus on the transfer of technology for mitigation and adaptation to developing countries. As with everything in the UNFCCC the process is rather formalised, with a Technology Executive Committee managing Technology Needs Assessments and a Climate Technology Centre and Network.

Finally, COP19 in 2013 set up the 'Warsaw Institutional Mechanism for Loss and Damage', a mechanism for considering how to provide developing countries with some form of funding to cover losses from extreme events generated by climate change. This is a challenging topic – not least because (as we have seen in Chapters 2 and 3) it is extremely difficult to attribute specific extreme events to climate change rather than climatic variability.

NATIONAL AND REGIONAL LEGISLATION AND FRAMEWORKS

The UNFCCC provides the international framework for countries to address climate change, but in practice actions are informed, facilitated and constrained by national or regional policies and legislation. Most developed and many developing countries now have some form of national legislation referring to climate change, with a very strong focus on reducing emissions of greenhouse gases. The first legislation specifically referring to climate change was passed in Japan in 1998 in the wake of the Kyoto Protocol. Most other legislation has been passed since 2008. Detailed assessments of national climate legislation in 66 countries are published by GLOBE International,[22] an international organisation comprising parliamentarians from over 80 countries.

Political frameworks take a variety of forms. Some are based on legislation passed by national and regional parliaments, and others are based on presidential executive orders; still others are informed by national five-year plans. All contain measures to reduce emissions of greenhouse gases, but most also include some reference to adaptation. There have been three major motivations for legislation.[23] Most legislation has been triggered by economic motivations, specifically relating to 'green growth' and energy security. Some has been influenced by a desire to demonstrate leadership (as with the Japanese legislation in 1998 and legislation in other countries which has been associated with COP meetings). Legislation in many developing countries has been motivated by perceived vulnerability to climate change, and in these countries adaptation tends to be the main focus of legislation. However, the delineation between 'adaptation' and 'managing climate risks' is not always clear, reflecting the close links between

22 www.globeinternational.org/about-globe. Nachmany, M., et al. (2014), *The GLOBE Climate Legislation Study: A Review of Climate Change Legislation in 66 Countries*. Fourth Edition. London: GLOBE International and the Grantham Research Institute, London School of Economics.

23 GLOBE International (2011), *GLOBE Climate Legislation Study*. Edited by Terry Townshend, Sam Fankhauser, Adam Matthews, Clement Feger, Jin Liu and Thais Narciso.

climate change adaptation and management of current climatic variability.

National legislation and regulations typically do different things in the areas of mitigation and adaptation. Those relating to mitigation tend to initiate actions (such as initiate an emissions trading scheme) or stipulate targets (such as defining targets for the use of renewable energy). Legislation and regulations relating to adaptation, however, are typically limited to requirements to undertake risk assessments, and usually delegate powers to local governments or agencies: this is because adaptation tends to be more diverse and local than energy policy. In 2013 GLOBE International produced a set of 13 'guiding principles' for incorporating climate change adaptation into legislation;[24] amongst other things, the principles emphasise the role of the private sector in supporting adaptation and the need for legal frameworks to be designed to allow organisations to be able to adapt.

The rest of this section summarises some of the key aspects of climate change legislation in a number of countries. It starts, however, with the European Union because this provides a framework for legislation in member states.

EUROPEAN UNION

The European Union influences actions within member states through Directives and Regulations, and the EU is itself a signatory of the Kyoto Protocol. The greatest effort has been directed so far on seeking to reduce emissions. The cornerstone of this has been the 'Climate and Energy Package' (CARE), adopted by the European Parliament in

24 www.globeinternational.org/the-legislative-principles-on-adaptation.

2009.[25] This set the so-called 20-20-20 targets, committing the EU to a 20% reduction in emissions by 2020 (relative to 1990), to raising the share of EU energy consumption produced from renewable resources to 20%, and to improving energy efficiency by 20%. It has been implemented through a series of directives relating to emissions trading (defining the European Union Emissions Trading System), energy efficiency, renewable obligations and the development of carbon capture and storage. Other directives setting energy efficiency standards for transport and specific types of equipment, along with the Biofuels Directive mandating an increased substitution of biofuels for fossil fuels, contribute to the targets. In 2011 the European Commission proposed a 'Roadmap for moving to a competitive low-carbon economy in 2050', which proposed greater targets for emissions reductions (40% by 2030, 60% by 2040 and 80% by 2050) and more specific sectoral targets. Although approved by the European Parliament, the Roadmap was blocked by Energy Ministers (specifically Poland) in 2012. The European Commission therefore presented a '2030 Framework' in January 2014, which is basically a revision of the 2050 Roadmap: it sets a target 40% reduction by 2030, but does not look further ahead.

The EU adopted an Adaptation Strategy in 2013, following a White Paper in 2009. Adaptation decisions are typically taken at national, regional and local levels, so rather than set specific targets and standards the EU's Adaptation Strategy has three aims. It promotes action by member states (encouraging states to adopt adaptation strategies – 15 had done so by January 2014), seeks to 'climate-proof' action at the EU level by mainstreaming adaptation into sectoral policies and ensuring Europe's infrastructure is made more

25 http://ec.europa.eu/clima/policies/brief/eu/index_en.htm.

resilient, and supports 'better informed decision-making' by making information and guidance available (through the Climate-ADAPT platform[26]). Support includes funding for building capacity in institutions in member states. 'Climate-proofing' includes incorporating the risks posed by climate change into agricultural and water management policies, for example, and also infrastructure building standards; the EU is currently considering how to incorporate climate change into the EN Eurocode standards for the design of buildings and infrastructure.

UK

The UK's flagship legislation relating to climate change is the 2008 Climate Change Act, which builds on earlier legislation referring to energy efficiency and is complemented by a raft of legislation applying to the energy system. The Act focuses on mitigation and the reduction in emissions. Central to the Act is a binding commitment to reduce UK emissions of greenhouse gases through a series of binding five-year carbon budgets. The UK Climate Change Act established the independent Committee on Climate Change (CCC), which sets the targets and budgets, and advises the government on measures to achieve targets. In 2009 the Scottish Parliament passed the Climate Change (Scotland) Act.

In its first report in 2008, the CCC proposed the long-term target for the UK to reduce emissions of greenhouse gases by 80% by 2050, relative to 1990,[27] and this is now embedded in both the UK and Scottish Climate Change Acts. This assessment was based on two considerations.

26 http://climate-adapt.eea.europa.eu/.

27 Committee on Climate Change (2008), *Building a low-carbon economy: the UK's contribution to tackling climate change.*

The first was an assessment of the reduction in global emissions necessary to achieve a target of limiting the rise in global mean temperature to 2°C above pre-industrial levels (1.4°C above current levels). More precisely, the CCC defined a reduction in emissions necessary to have a 50% chance of keeping the rise in temperature below 2°C, and very little chance of temperature rising above 4°C. This formulation recognises the fact that it is impossible to predict exactly the rise in temperature for a given pathway of future emissions. Analysis for the CCC evaluated many potential emissions pathways, and the pathways which achieved the target standards had emissions declining after 2016 at 3-4% per year (Figure 4.3a and b). By 2050, annual global emissions would need to be approximately 50% below 1990 levels. The second consideration was that the UK should reduce its emissions by more than the global average, to reflect its historical contribution to increased greenhouse gas concentrations in the atmosphere and to move towards a more equal global per capita rate of emissions; the UK should therefore reduce its emissions not by 50%, but by 80%.

Figure 4.3c shows the UK's emissions in 2006, by sector, and the target emissions in 2050. The CCC argued that the reduction of 80% (or 77% from 2006) was technically feasible. Figure 4.3d shows the UK's total emissions in 1990 and 2006, together with the target emissions under the first four five-year carbon budgets.[28] Mitigation measures to achieve the carbon budgets are defined in the UK's Carbon Plan, presented to Parliament in 2012. It is likely that the first three five-year budgets will be met, but the fourth five-year budget (2023-2027) is likely to be exceeded without additional measures.

28 The five-year budgets actually define five-year totals rather than annual values, but they have been expressed in annual terms here for comparison with the 1990 and 2006 totals.

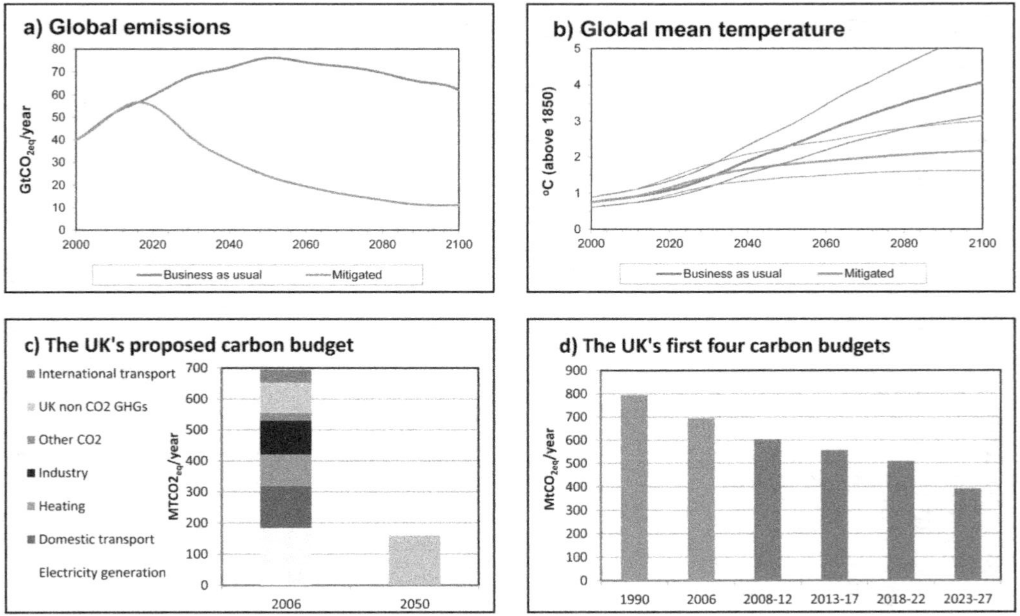

Figure 4.3 **The UK's climate mitigation policy: (a) Global emissions under business-as-usual and a pathway that achieves the temperature target; (b) global mean surface temperature; (c) UK emissions in 2006 and 2050; and (d) the first four UK five-year carbon budgets, together with emissions in 1990 and 2006**

Adaptation was added to the UK Climate Change Act at a late stage. The Act created the Adaptation Sub-Committee of the CCC, with a remit to advise on and monitor progress towards adaptation, and required the UK government to conduct climate change risk assessments every five years (the first CCRA was published in 2012,[29] as outlined in Chapter 3). Adaptation is a responsibility of the devolved administrations within the UK, so the UK Act requires these administrations to develop their adaptation strategies (the Climate Change (Scotland) Act also requires the development of a Scottish adaptation strategy). The strategies differ between the administrations because priorities and context are different.

The first UK National Adaptation Programme was published in July 2013.[30] In principle it deals with matters reserved to the UK government as well as issues relevant only to England, but in practice it focuses on adaptation within England. It defined 31 objectives (Table 4.5), each with a series of specific actions. The Welsh Assembly published its Climate Change Strategy for Wales[31] in 2010, which includes an Adaptation Framework: this focuses on building the evidence base, mainstreaming adaptation into decision-making, and communicating the need for adaptation; the Adaptation Delivery Plan defines 24 actions (Table 4.6). The Northern Ireland Climate Change Adaptation Plan was submitted to the Northern Ireland Assembly in January 2014.[32] It identifies four priority areas – flooding, water, the natural environment and agriculture and forestry – with between five and seven specific high-

29 Defra (2012), *UK Climate Change Risk Assessment: Government Report*. HM Government.

30 Defra (2013), *The National Adaptation Programme. Making the Country Resilient to a Changing Climate*. Defra.

31 Welsh Assembly Government (2010), *Climate Change Strategy for Wales*.

32 Department of the Environment Northern Ireland (2014), *Northern Ireland Climate Change Adaptation Programme*.

Table 4.5 Specific objectives in the National Adaptation Plan for England

Policy area	Objective
Built Environment	1: Work to reduce the threat of flooding and coastal erosion
	2: Provide clear local planning framework to deliver sustainable development
	3: Help business and industry in the sector to access skills and knowledge
	4: Ensure that investors and developers have appropriate tools to support adaptation
	5: Increase resilience of homes and buildings
	6: Build understanding of long-term implications of climate change for location and resilience of population centres
Infrastructure	7: Ensure infrastructure is located and maintained to be resilient to climate change
	8: Develop regulatory frameworks to support and promote a resilient infrastructure sector
	9: Better understand vulnerabilities facing 'local' infrastructure
	10: Develop understanding in managing interconnected services
Health	11: Reduce risk of death and illness associated with severe weather events and climate change
	12: Promote resilience within the health and social care system
	13: Minimise impact of climate change on vulnerable groups by strengthening their resilience
	14: Promote and strengthen community resilience to severe weather events
Agriculture and forestry	15: Increase resilience of agriculture by managing impact of volatility in rainfall

	16: Increase resilience of forestry sector by increasing level of management
	17: Increase resilience to pests and diseases
	18: Embed adaptation into agriculture, forestry and horticulture research programmes
Natural environment	19: Build resilience of wildlife, habitats and ecosystems to climate change
	20: Take action to help wildlife, habitats and ecosystems transition through change
	21: Promote widespread uptake of adaptation measures that benefit or do not adversely affect the natural environment
	22: Improve the evidence base and understanding of decision-makers and others
Business	23: Raise awareness and understanding about climate change risks
	24: Increase extent to which businesses consider climate change impacts
	25: Raise awareness of domestic and international opportunities
	26: Help businesses better manage risks to supply chains
	27: Undertake research to increase understanding of climate change impacts on growth and the economy
Local government	28: Raise and maintain profile of adaptation and embed resilience
	29: Support local government to build credible business case for action
	30: Ensure policy framework supports councils to increase resilience
	31: Support activities to allow councils to make local commitments to address local challenges and opportunities

Table 4.6 Specific objectives in the Welsh Adaptation Delivery Plan

Theme		Building evidence base	Mainstreaming	Communicating
Strategic actions	1: Build consideration of climate change impacts into Welsh Assembly business planning		X	
	2: Assess risks and opportunities	X		
	3: Integrate assessment of risks/opportunities and development of adaptation plans into all public bodies		X	
	4: Increase awareness and understanding of impacts	X		X
	5: Target resources to support adaptation			X
Natural environment	6: Develop understanding of impacts	X	X	X
	7: Ensure that marine spatial planning considers impacts	X	X	
	8: Support land managers to adapt		X	X
	9: Improve woodland resilience		X	
Infrastructure	10: Consider impacts in sustainable infrastructure development and regeneration		X	
	11: Publish flood and erosion risk management strategies		X	
	12: Support spatial planning to consider impacts		X	

	13: Improve resilience of water infrastructure		X	
	14: Identify opportunities to support adaptation in built environment		X	
	15: Review resilience of transport network and develop programme		X	
	16: Review Shoreline Management Plans		X	
	17: Explore consequences for the historic environment	X		
Communities	18: Build adaptation into local authority processes		X	
	19: Inform community and third-sector organisations and help them adapt	X	X	X
	20: Raise awareness of flood risk			X
	21: Progress partnership with community in Uganda			
Health	22: Implement heatwave plan		X	X
	23: Implement adaptation action plan for health		X	X
Business and tourism	24: Provide adaptation advice to business			X

level actions (Table 4.7). The draft Scottish Climate Change Adaptation Programme was published in 2013;[33] it has nine specific objectives in three focus areas (Table 4.8).

The UK Act also introduces reporting powers to require public service organisations (including public agencies and

33 Scottish Government (2013), *Climate Ready Scotland: Scottish Climate Change Adaptation Programme*.

Table 4.7 Themes and high-level actions in the Northern Ireland Climate Change Adaptation Programme

Theme	High-level action
Flooding	Develop and implement sustainable flood risk management strategies
	Build resilience into infrastructure
	Identify and implement measures to reduce impact of significant risks
	Develop stakeholder understanding
	Identify future flood risks to built heritage
Natural environment	Build resilience into natural environment
	Explore research opportunities
	Develop new monitoring programmes
	Support land managers in building resilient natural environment
	Protect and enhance the management of habitats and species
	Increase awareness
	Incorporate climate change into marine planning
Water	Implement water management measures
	Improve resilience of water resource infrastructure
	Improve water efficiency
	Implement water monitoring programmes
	Maintain and enhance quality of water
	Raise awareness
Agriculture and forestry	Increase awareness of adaptation
	Identify climate resilient grasses and crops
	Identify measures to support adaptation
	Improve resilience of woodland
	Encourage land managers to reduce risk of wildfires

Table 4.8 Specific objectives in the draft National Adaptation Plan for Scotland

Theme	
Natural environment	N1: Understand the impacts of climate change on the natural environment
	N2: Support a healthy and diverse natural environment with capacity to adapt
	N3: Sustain and enhance the benefits, goods and services that the environment provides
Buildings and infrastructure networks	B1: Understand the impacts of climate change on buildings and infrastructure
	B2: Provide the knowledge and tools to manage impacts
	B3: Increase the resilience of buildings and networks
Society	S1: Understand the impacts of climate change on people, homes and communities
	S2: Increase awareness of the impacts of climate change
	S3: Support health services and emergency responders

statutory undertakings – such as energy, water and transport utilities) to undertake risk assessments and report on their adaptation plans. By mid-2013, nearly 90 organisations in England had provided reports.

USA

The USA has not ratified the Kyoto Protocol, and has no specific climate change legislation at the Federal level;[34] indeed, such legislation has proved to be highly controversial.

34 Although there is legislation from 2013 preventing US airlines from joining the EU Emissions Trading System.

Table 4.9 The US Federal adaptation strategy: preparing the United States for the impacts of climate change

Building stronger and safer communities and infrastructure	Direct agencies to support climate-resilient investment
	Establish a State, Local and Tribal Leaders Task Force on Climate Preparedness
	Support communities as they prepare for climate impacts
	Boost the resilience of buildings and infrastructure
	Rebuild and learn from Hurricane Sandy
Protecting the economy and natural resources	Identify vulnerabilities of key sectors to climate change
	Promote resilience in the health sector
	Promote insurance leadership for climate safety
	Conserve land and water resources
	Maintain agricultural sustainability
	Manage drought
	Reduce wildfire risks
	Prepare for future floods
Using sound science to manage climate impacts	Develop actionable climate science to develop tools
	Assess climate change impacts
	Launch open climate data initiative
	Provide toolkit for climate resilience

However, there are many pieces of legislation mandating energy efficiency standards, and since 2009 the Environmental Protection Agency (EPA) has been able to regulate CO_2 as a pollutant under the 1963 Clean Air Act. This has been controversial, but it provides a means of curbing emissions without setting specific national targets.

In 2013 President Obama launched the US Federal Climate Action Plan,[35] which basically seeks to reduce emissions and encourage adaptation through a series of Executive Orders. Table 4.9 summarises the key actions in the area of adaptation; a Federal Interagency Climate Change Adaptation Task Force seeks to coordinate activities of Federal agencies.

Several individual states (including California, Arizona, Connecticut, Massachusetts and Pennsylvania) have enacted legislation specifically addressing climate change. Some of this legislation defines emissions targets, and most introduce measures to encourage renewable sources of energy or increase energy efficiency. Most states have also developed some form of adaptation strategy: California's is a good example.[36] As within the UK, these adaptation strategies focus on issues of greatest local importance.

SOUTH AFRICA

Whilst most developed country's national policies towards climate change focus on mitigation, South Africa's National Climate Change Response Policy, agreed in 2011,[37] places a greater emphasis on adaptation and risk management. It places this as its first objective, whilst making a 'fair contribution' to the global effort to stabilise greenhouse gas concentrations is placed second. Eight priority areas are identified for the development of adaptation strategies, each with a set of specific objectives. These are: water, agriculture

35 Executive Office of the President (2013), *The President's Climate Action Plan*. The White House, Washington.

36 Moser, S., et al. (2012), *Our Changing Climate 2012: Vulnerability and Adaptation to the Increasing Risks from Climate Change in California*. A Summary Report on the Third Assessment from the California Climate Change Center.

37 Government of South Africa (2013), *National Climate Change Response White Paper*.

and commercial forestry, health, biodiversity and ecosystems, urban settlements, rural settlements, coastal settlements and disaster risk reduction.

OVERVIEW

This section has provided a snapshot of climate policy in three countries – the UK, USA and South Africa. Many countries have some form of legislation relating to climate change, and even more have specific legislation addressing energy efficiency and sources. Specific climate change legislation has proved controversial in many countries. The USA is a notable example, but Australian federal climate change policy has fluctuated considerably; legislation to establish mechanisms to reduce greenhouse gas emissions was finally passed in 2011, only to be repealed in 2013.

Most countries (and states within federal countries) have developed some form of adaptation strategy, and the strategies for the parts of the UK are presented in this Chapter. Although the precise priorities of national and regional strategies vary with local circumstances and priorities, there are a number of common themes.

Adaptation strategies usually include measures to raise awareness, build capacity and make specific interventions (adaptation actions). Most strategies emphasise that adaptation is best done in partnership between governments, public agencies, the private sector and 'third-sector' organisations. Adaptation frameworks therefore include many 'enabling' elements. Finally, many strategies which are developed to address climate change actually include measures which could be regarded as best practice when dealing with current climatic variability.

REPORTING ON CLIMATE RISKS: REGULATION AND VOLUNTARY REPORTING

Regulations in many countries require organisations to report on their energy use and the energy efficiency of their facilities. However, there are far fewer requirements on organisations to report on how they deal with climate risks or are planning to adapt to climate change. Such information would potentially be useful in a number of ways. It allows governments to monitor progress towards meeting adaptation objectives set out in strategies. It allows investors to assess how organisations are dealing with risks to their business and operations. Thirdly, it enables organisations to benchmark themselves against other organisations. Reporting on climate risks can be required by regulators, or can be provided voluntarily.

REGULATORY REQUIREMENTS FOR REPORTING

Under the UK's Climate Change Act, public bodies and statutory undertakings are required to submit reports on their assessment of the risks posed by climate change and their adaptation strategies. The statutory undertakings are mostly private sector companies running key public utilities, including water, energy and transport. There are no requirements for other private sector organisations to report on their climate risks. Under the terms of the UK's Financial Conduct Authority, companies listed on the London Stock Exchange must present descriptions of principle risks and uncertainties in their annual financial reports, but there is no obligation to report on climate change risks specifically.

In the United States, however, the Securities and Exchange Commission (SEC) – which regulates the US securities markets – has since 2010 required companies to report on their exposure to climate risks.[38] This follows a campaign by investors and environmental groups for the SEC to make it clear to companies that they needed explicitly to consider climate risks under their reporting obligations. Despite being required to report on risks that would be 'material' to a shareholder's investment decisions, surveys found that few did;[39] in 2008 only 24% of S&P 500 companies identified climate change risks, and only 5.5% discussed management strategies for the risk. The SEC guidance identifies four areas where climate change could trigger disclosure requirements: the impact of legislation and regulation, the impact of international accords, the indirect consequences of regulation or business trends (e.g. on demand), and the physical impacts of climate change. Reporting has improved slightly since 2010 (59% of S&P 500 companies reported in 2013), although it has been argued that the majority of reporting does not fully meet SEC requirements, is vague, and tends to focus on regulatory risk.[40] Specific guidance for companies on disclosure was produced in 2012,[41] including check-lists of relevant questions by industry sector.

Also in the United States, the National Association of Insurance Commissioners (NAIC: the association of state insurance regulators) conducts the Insurer Climate Risk Disclosure Survey. This was undertaken in 2010 and, most comprehensively, in 2012. The survey involves just eight questions, focusing on how companies incorporate climate

38 Commission Guidance Regarding Disclosure Related to Climate Change (Feb. 2, 2010), available at www.sec.gov/rules/interp/2010/33-9106.pdf.

39 CERES (2009), *Reclaiming transparency in a changing climate: trends in climate risk disclosure by the S&P 500 from 1995 to the present.* www.ceres.org.

40 CERES (2014), *Cool response: the SEC and corporate climate change reporting.* www.ceres.org.

41 Calvert Investments, Oxfam and CERES (2012), *Physical Risks from Climate Change: A guide for companies and investors on disclosure and management of climate impacts.*

risks into their plans and engage with stakeholders. A review of the 2012 survey results[42] showed relatively little appreciation of the potential effects of climate change and that very few companies had specific comprehensive strategies to cope with climate change risks.[43]

VOLUNTARY REPORTING OF CLIMATE CHANGE RISKS

Several international initiatives have developed to provide consistent information on company performance and behaviour in relation to environmental issues. The Global Reporting Initiative (GRI)[44] was established in 1997 as an international non-profit organisation with the aim of helping companies collate and publish standardised information and data against a range of sustainability indicators. It has developed a Sustainability Reporting Framework which allows companies to measure, understand and communicate performance and impact in four key areas: economic, environmental, social and governance. The framework comprises a large number of indicators, including greenhouse gas emissions and – within the economic dimension – the financial implications, risks and opportunities due to climate change. Quantitative data are required for emissions, but narrative information is used to describe climate risks. The GRI essentially provides a means for companies to produce their own sustainability reports in a consistent way to allow them to compare and benchmark themselves and provide information to investors. It does not synthesise or summarise company reports.

42 CERES (2013), *Insurer climate risk disclosure survey: 2012 findings and recommendations*. www.ceres.org.

43 The CERES summaries of responses to SEC and NAIC reporting requests provide useful benchmarking information for companies.

44 www.globalreporting.org.

The CDP (previously Carbon Disclosure Project)[45] is another international non-profit organisation, but its remit is to both collect company climate change (and now water and forest) data and to publish analytical reports highlighting trends and developments. It has around 5000 member companies worldwide, who each submit consistent data annually. The annual climate change information request seeks data and information not only on company governance, but also on emissions and both the risks and opportunities posed by climate change. Its 2013 Supply Chain report[46] found that 75% of companies reported that they believed their supply chain was subject to disruption or increased costs due to climate change. Its 2013 report on FTSE350 companies' understanding of the international risks of climate change[47] showed that whilst most companies identified risks and opportunities, most were focused on direct, short-term risks.

The Climate Disclosure Standards Board (CDSB)[48] is another non-profit organisation, established specifically to promote and advance climate change-related disclosure in mainstream company reports. It has developed a global framework for reporting on climate change – covering not only emissions but also risks – and differs from the GRI in the sense that it is explicitly seeking to 'mainstream' climate change information rather than present it in a separate 'sustainability report'.

CONCLUSION

This chapter has examined the political and regulatory context for climate change. This political context has been

45 www.cdp.net.

46 CDP (2013), *CDP Supply Chain Report 2012-2013* CDP.

47 CDP (2013), *Are UK companies prepared for the international impacts of climate change?* FTSE 350 Climate Change Report CDP.

48 www.cdsb.net.

in many cases highly controversial, but international and national legal frameworks have been established to support policies to address the causes and consequences of climate change. Most of these frameworks have focused on reducing the emissions of greenhouse gases, and the UNFCCC process is central to this. However, many countries have also established policies for supporting adaptation to climate change risks.

There are several general summary points.

- International policy is aiming to limit the increase in global mean temperature to 2°C above pre-industrial levels (or about 1.4°C above current levels). This is challenging, but technically feasible; whether or not it is politically and economically feasible is another question. However, even if the rise in temperature were to be limited to 2°C, there would still be some impacts from the changing climate over the next few decades.

- The uneven adoption of legislation and regulations (and the occasional changes in direction) show that regulatory risk is potentially very important. This is most obviously the case on the mitigation side (where the cost of energy may be very influenced by regulatory uncertainty), but is also relevant when developing measures to adapt to climate change risks.

- Whilst national and regional adaptation strategies are geared towards local needs, a common theme is the need for partnership between the public and private sectors. The actions of private sector organisations will be affected by public policy towards adaptation, and many public goals can only be delivered through the private sector.

- There is an emerging demand for consistent and transparent information on how organisations are exposed to, and cope with, the risks and opportunities posed by climate change.

5 Assessing Climate Change Risks and Opportunities

INTRODUCTION: ADDRESSING RISKS AND OPPORTUNITIES THROUGH RISK MANAGEMENT PROCESSES

It is clear from the previous chapters that climate change has the potential to pose risks to, and create opportunities for, an organisation. These risks and opportunities are best considered within a *risk management framework,* which defines responsibilities for dealing with risk and establishes a *risk management process* for addressing risk. Figure 5.1 shows the general risk management framework and risk management process under the international standard ISO31000:2009, which has been embedded in many national standards.

Several attempts have been made to construct risk management processes specifically to address climate change. Figure 5.2 shows

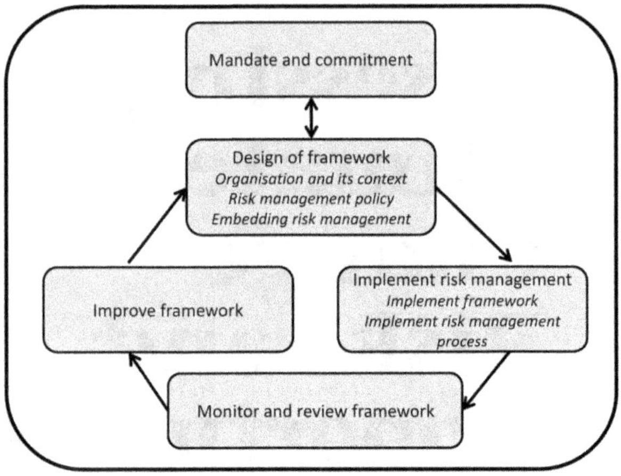

Figure 5.1a Risk management under ISO31000:2009 – Risk management framework

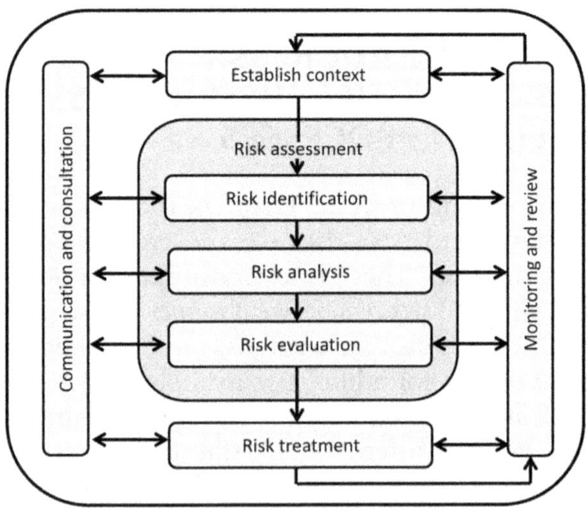

Figure 5.1b Risk management under ISO31000:2009 – Risk management process

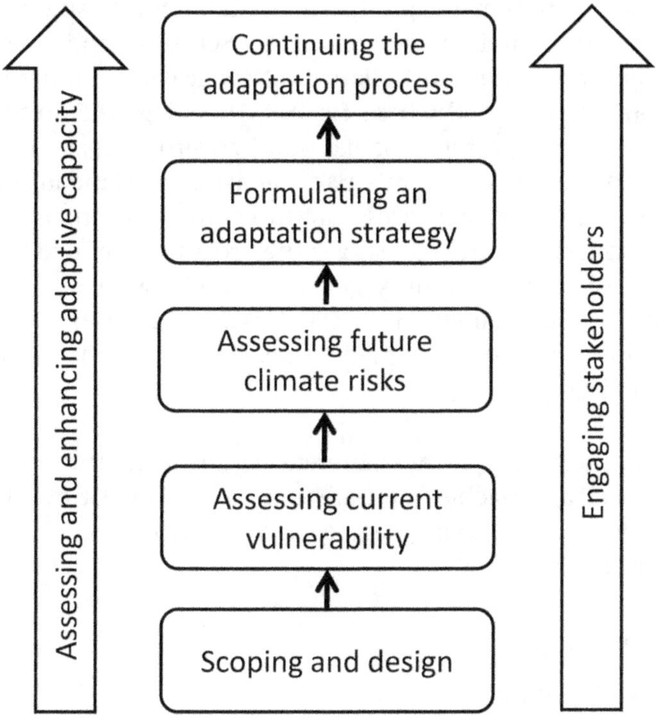

Figure 5.2 The UNDP Adaptation Policy Framework

the approach defined in the United Nations Development Programme (UNDP) Adaptation Policy Framework,[1] aimed specifically at incorporating adaptation to climate change into development projects. More models of process, targeted specifically at individual organisations considering adapting to climate change, are shown in Figure 5.3. Figure 5.3a shows the processes recommended in the Australian guide on impacts

1 Lim, B., et al. (2004), *Adaptation Policy Frameworks for Climate Change: Developing Strategies, Policies and Measures*. Cambridge University Press for UNDP.

and risk management produced in 2006 for business and government.[2] This looks very similar to ISO31000, and indeed it is explicitly based on Australian and New Zealand Standard AS/NZ 4360 which was the basis for ISO31000. Figure 5.3b shows a process proposed for Canadian local governments,[3] which is based on Canadian national risk management standards, and Figure 5.3c shows a process produced for Californian local communities.[4] Figure 5.3d shows the process as defined in the UKCIP risk and uncertainty guidance,[5] and Figure 5.3e shows the process implemented in the UKCIP Adaptation Wizard tool,[6] an online tool designed to help organisations adapt to climate change.

Whilst there are some differences between the various conceptual approaches, all share some common features. They all include, with varying degrees of distinction, the following components:

- scoping;
- risk identification;
- risk analysis;
- risk evaluation;
- risk treatment;
- monitoring and review.

2 Australian Greenhouse Office (2006), *Climate Change Impacts and Risk Management: A Guide for Business and Government*. Commonwealth of Australia.

3 Black, R., et al. (2010), *Adapting to Climate Change: A Risk-Based Guide for Local Governments*. Canada.

4 California Emergency Management Agency (2012), *California Adaptation Planning Guide: Planning for Adaptive Communities*. CEMA.

5 Willows, R. and Connell, R. (eds) (2003), *Climate Adaptation: Risk, Uncertainty and Decision-Making*. UKCIP, Oxford.

6 www.ukcip.org.uk/wizard.

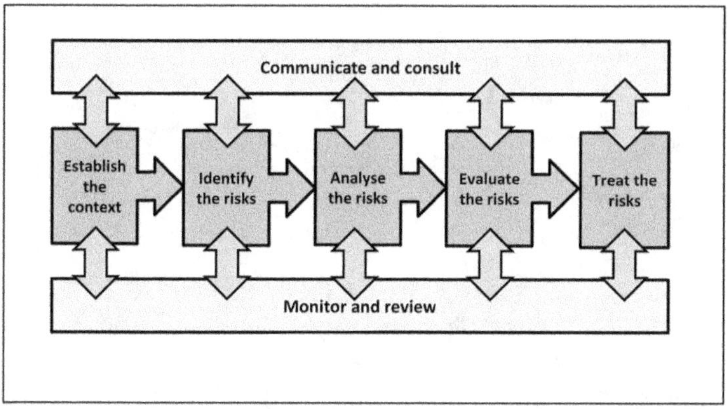

Figure 5.3a Five frameworks for climate change risk assessment and management: Australia

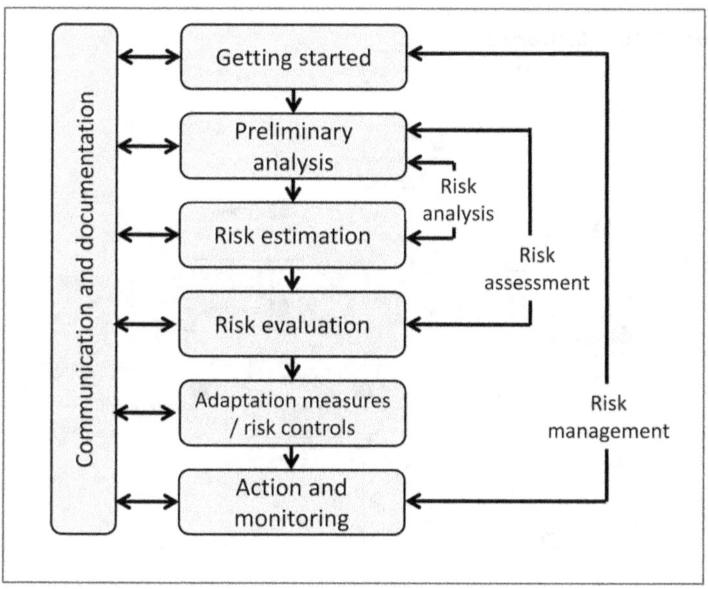

Figure 5.3b Canada

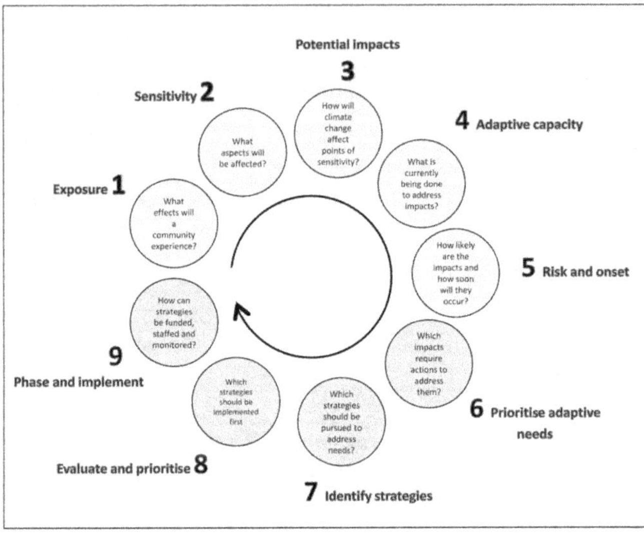

Figure 5.3c California

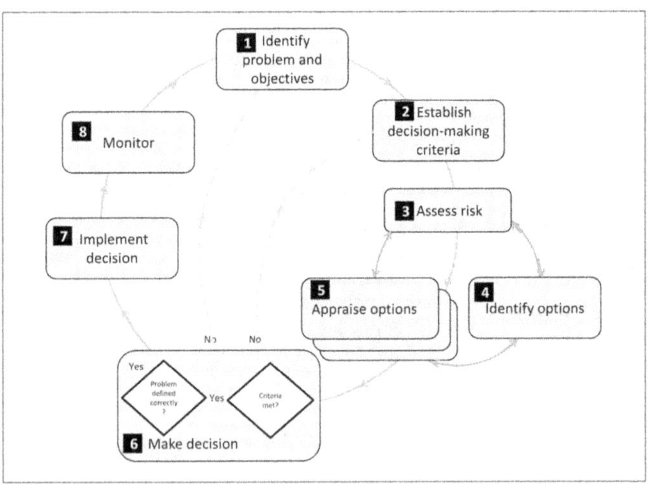

Figure 5.3d The UKCIP framework

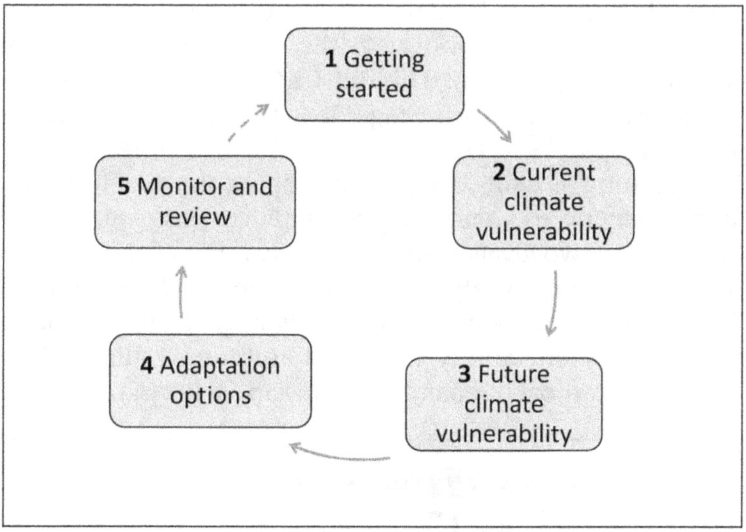

Figure 5.3e The UKCIP online Adaptation Wizard

These processes fall into two groups. The first set – from scoping through to risk evaluation – deals with the assessment of risk, and is covered in this chapter. The second set covers the selection and implementation of measures to cope with any climate change risks, and is discussed in Chapter 6.

RISK ASSESSMENT

The first stage in any risk assessment is a scoping exercise, designed to define the boundaries and criteria for the assessment. The next stages seek to identify potential risks (and opportunities), analyse or quantify those risks, and finally evaluate them both against each other and in terms of the organisation's attitudes towards risk. The organisation of this section largely follows the components of a risk assessment as specified in the ISO31000 and Australian approaches (which are virtually identical to the corresponding parts of the Canadian approach). The full UKCIP framework defines risk assessment within its Stage 3, with sub-components as shown in Figure 5.4. This proposes a tiered approach. Tier 1 'risk screening' is equivalent to the scoping stage, whilst Tiers 2 and 3 (generic and detailed quantitative risk assessment) are equivalent to two levels within risk analysis; risk prioritisation is equivalent to risk evaluation. The online UKCIP Adaptation Wizard represents risk assessment in its stages 2 and 3 looking separately at vulnerability to current and future climate. The Californian approach separates out specific actions differently.

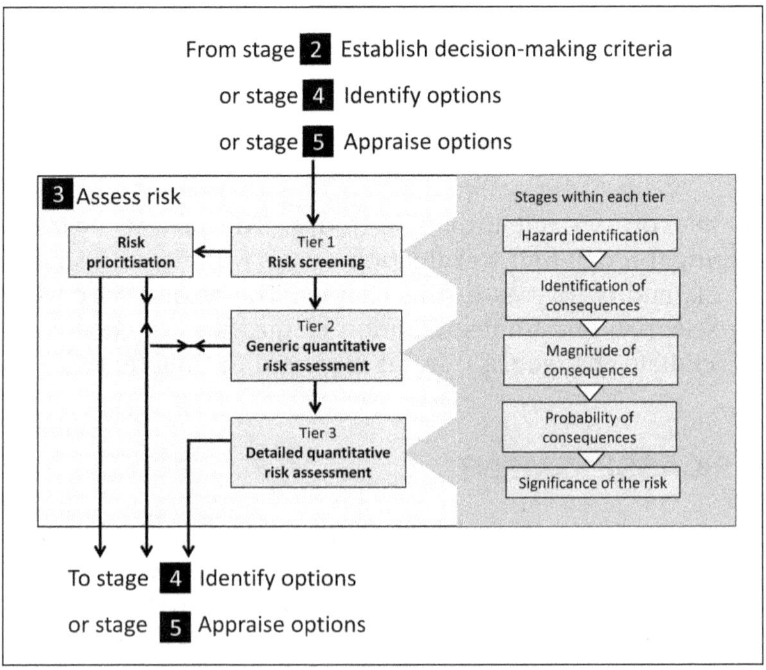

Figure 5.4 Stages in risk assessment in the full UKCIP framework (Willows and Connell, 2003)

However, all include, explicitly or implicitly, the same set of activities and issues outlined below.

SCOPING: ESTABLISHING THE CONTEXT

The scoping exercise sets out the context for a risk assessment. There are three components to this context: organisational, conceptual and practical. The *organisational* issues include defining which parts of an organisation need to be involved, which external stakeholders need to be consulted, and who is empowered to make decisions. These decisions relate not only to choosing adaptation options (discussed in Chapter 6), but also to the scope and boundaries of the risk assessment. *Conceptual* issues include recognising and defining an organisation's 'risk appetite' and its commitment either to innovation or adherence to established best practice. Risk appetite describes the level of risk that an organisation is prepared to accept, and will depend on a range of factors including organisational culture and business and regulatory context. These are all very generic topics, relevant to any type of risk assessment.

The *practical* issues which are more specific to climate change include the definition of the time horizon of interest, understanding of the 'coping range' of an organisation, the identification of components of an organisation which need to be involved, and the specification of the level of detail. Time horizon is important because the time scales over which climate change is likely to manifest itself may be longer than those over which an organisation plans – and there may be important differences between strategic long-term planning and operational planning over the shorter term. The 'coping range' describes the variability in conditions that an organisation can readily cope with, for example through existing contingency plans or storages. The identification of the components of an organisation which need to be involved requires an appreciation of the degree to which parts of an organisation are sensitive to weather and extreme events. Finally, the specification of the level of detail required defines the approaches which should be used in later stages. For example, a high-level long-term strategic assessment is unlikely to need the use of detailed climate scenarios covering a wide range of weather variables, whilst a more targeted assessment might require considerable detail.

This scoping stage is equivalent to Stages 1 and 2 in the full UKCIP framework, and is implicitly a 'Step 0' in the Californian approach.

RISK IDENTIFICATION

This stage seeks to identify the climate risks to which an organisation is exposed and the opportunities provided by climate change, before selecting a subset of these risks and opportunities for more detailed investigation in the next 'risk analysis' stage.

Chapter 3 demonstrated that vulnerability to climate change was dependent on exposure to climate, sensitivity to change, and the ability to cope or adapt. This is encapsulated in Figure 5.5,[7] which shows how exposure and sensitivity combine to produce a potential impact. The extent to which this impact is a problem depends on the organisation's ability to cope with the potential impact *with its standard set of procedures and approaches*. If it cannot cope – or is not sure if it can cope

7 Preston, B.L. and Stafford-Smith, M. (2009), *Framing Vulnerability and Adaptive Capacity Assessment*. CSIRO Climate Adaptation Flagship Working Paper, CSIRO, Aspendale and Canberra, Australia.

– then the organisation is potentially vulnerable to climate change. If the likelihood of the change occuring is high, then the organisation is exposed to climate change risk; if the likelihood is low, then the risk is low even if the vulnerability is high. The Californian framework explicitly separates out exposure, sensitivity, impact and ability to cope as its stages one to four. A Tier 1 risk screening in the full UKCIP framework is equivalent to risk identification.

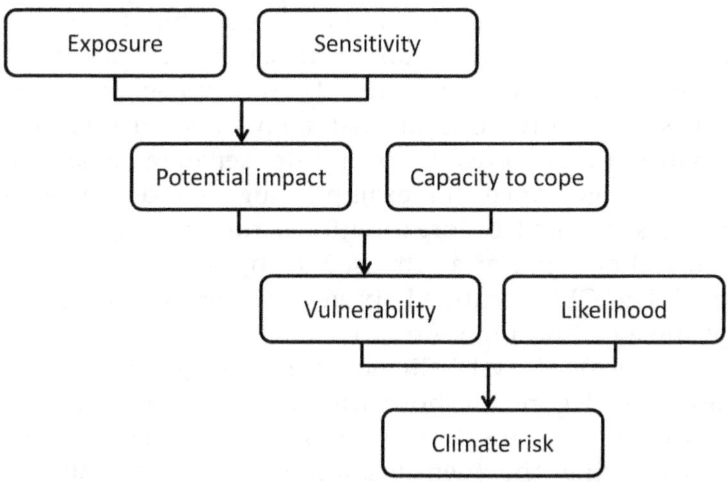

Figure 5.5 Climate risk as a function of exposure, sensitivity, capacity to cope and vulnerability (Preston and Stafford-Smith, 2009)

In this risk identification stage, the assessments of exposure, sensitivity to change, impact and likelihood will be based largely on some form of expert judgement or knowledge of the organisation's activities. The assessments may be informed by some high-level climate scenarios in a qualitative sense, if only to put some bounds on the types of change that might be anticipated. More usually, risk identification starts by looking at potential vulnerabilities and working back to see how they relate to weather and climate.

It is important that a robust process is used to identify risks to an organisation. There are two key components to this, both described in the various risk assessment frameworks.

The first is to take a structured approach to the search for potential risks and opportunities. A structured approach breaks the issues facing an organisation into a number of key topic areas or elements. These key elements could be based on any concept that allows an organisation's activities to be broken down into separate areas. The Australian guidance, for example, suggests key elements can be structured by organisational function, geographic area, technologies or assets used, or by service or product types. The CBI[8] identified six generic business areas, and the BACLIAT Business Areas Climate Impacts Assessment Tool,[9] developed by UKCIP, identifies a different set of six (Table 5.1). Figure 5.6 shows the 'risk disk'[10] characterising areas of impact, and Figure 3.8 presents yet another way of structuring the identification of risks. The Business Resilience Health Check Tool,[11] developed in the UK, is an online preliminary risk identification tool aimed primarily at small businesses; it guides users through seven business areas (the BACLIAT areas plus 'insurance').

---

8 CBI (2009), *Future Proof: Preparing your Business for a Changing Climate*, CBI London. CBI (2012), *Whatever the Weather: Managing the Risks from a Changing Climate*, CBI London.

9 Environment Agency (2013), *BACLIAT: Business Areas Climate Impacts Assessment Tool*. Climate Ready, Environment Agency, Bristol, UK.

10 Sussman, F.G. and Freed, J.R. (2008), *Adapting to Climate Change: A Business Approach*. Pew Center on Global Climate Change, Arlington, Virginia.

11 www.businessresiliencehealthcheck.co.uk.

Table 5.1 Generic business areas for climate change risk identification

BACLIAT	CBI
Logistics: vulnerability of supply chain, utilities and transport infrastructure	**Supply chains**: supply routes and sources for raw materials
Finance: implications for investment, insurance and stakeholder reputation	**Assets**: potential impact to the organisation's key infrastructure
Markets: changing demand for goods and services	**Operations**: possible impact on the organisation's core functions
Process: impacts on production processes and service delivery	**Markets**: resilience of distribution networks and potential shifts in market and customer behaviour
People: implications for workforce, customers and changing lifestyles	**Regulatory compliance**: regulatory frameworks governing firms' operations
Premises: impacts on building design, construction, maintenance and facilities management	**Reputation**: impact of climate risk on public perception of organisation ('licence to operate') and investor confidence

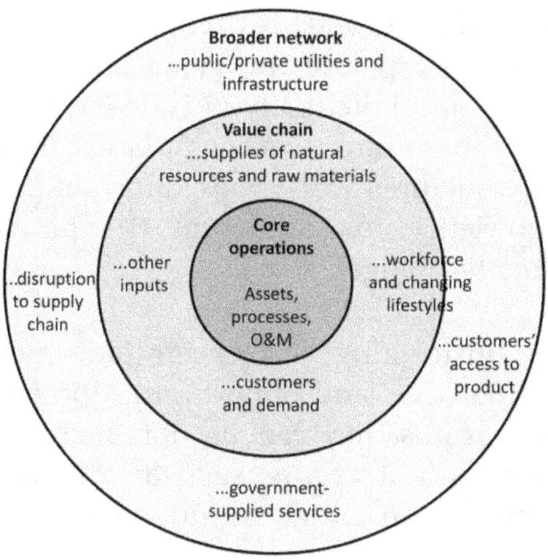

Figure 5.6 The 'risk disk': identifying business sectors exposed to climate change (Sussman and Freed, 2008)

Another example of a structured approach to risk identification is shown in Figure 5.7.[12] This 'risk screening' approach essentially asks three questions:

- is climate important to business risk?
- is there an immediate threat or are long-term assets, investments or decisions being locked into place?
- is a high value at stake if a wrong decision is made?

The screening results in three categories of outcome, distinguishing between potential significant risks that may need to be managed in the short term, potential threats that need to be monitored and reassessed over time, and risks that do not appear to be significant.

Regardless of the structure used for the identification of potential risks, the second key component to a robust evaluation is to incorporate the views and perspectives of stakeholders in – and potentially outside – the organisation. All the risk assessment frameworks emphasise the importance of conducting structured workshops, aimed at drawing upon employee knowledge to brainstorm the potential future impacts of climate change.

Aside from brainstorming and expert judgement, insights into the potential risks and opportunities posed by climate change to an organisation can be obtained by examining experience from recent events, such as an extreme storm, flood or heatwave, in order to understand vulnerability

12 Sussman, F.G. and Freed, J.R. (2008), *Adapting to Climate Change: A Business Approach*. Pew Center on Global Climate Change, Arlington, Virginia.

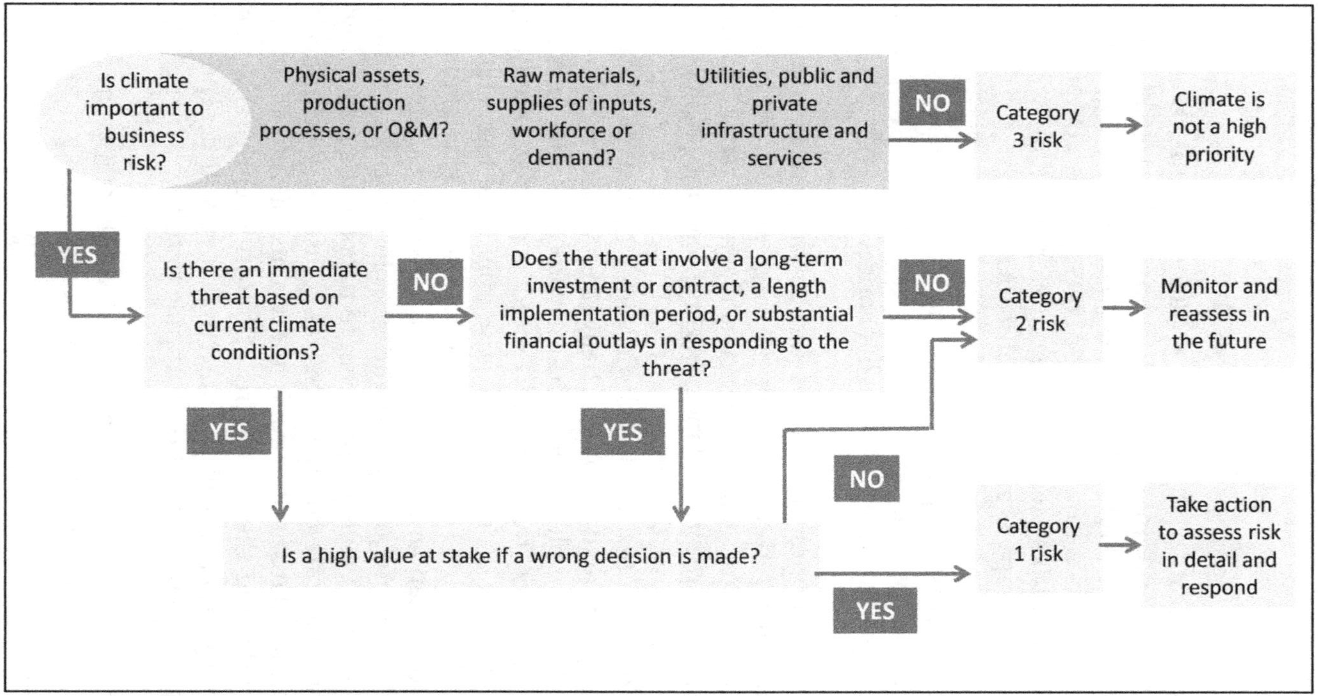

Figure 5.7 Screening for climate risk (Sussman and Freed, 2008)

to current weather risks. A key challenge here is to collate relevant evidence in a systematic manner. The UKCIP Local Climate Impacts Profile (LCLIP)[13] is a simple tool designed to collate and summarise information. It essentially comprises a spreadsheet with systematic categories for describing weather events, their impacts and their consequences.

A variant on this approach is to undertake a 'stress-test', to explore the consequences for the organisation of a specific hypothetical event or sequence of changes.

The risk identification stage is the opportunity to identify critical impact thresholds for an organisation. A critical threshold is the point at which sudden or rapid change occurs, or where circumstances begin to become challenging (or beneficial) to the organisation. There are two types of threshold. The first are based on some physical value, such as a temperature threshold or change in productivity of a key input. For example, some industrial processes may become inefficient once temperatures exceed a defined threshold, or demand for a product may arise only once temperatures pass a threshold. The second type of threshold is defined in terms of a standard of service, expressed for example as a failure rate or frequency of occurrence. In this case, a critical change occurs where events begin to happen 'too frequently' or 'not frequently enough'. This, of course, may be difficult to identify statistically unless the events of interest occur frequently, and the definitions of 'too frequently' and 'not frequently enough' will be subjective; they will depend, for example, on an organisation's attitude to risk.

Finally, it is important to remember that risks may come from many directions. They include not only the physical risks

13 www.ukcip.org.uk/wizard/current-climate-vulnerability/lclip/.

posed by climate change to an organisation, its supply chain and its customers, but also the risks presented by changes to (or uncertainty in) regulatory frameworks relating to climate change.

RISK ANALYSIS

The risk analysis stage involves the quantification of the potential risks defined in the risk identification stage. This quantification can take several different forms and operate at different levels of detail, depending on the potential significance of the risk and the resources available. The quantification can be iterative, with an initial quantification followed up by more detailed investigations if deemed necessary. Within the full UKCIP framework, these are 'Tier 2' and 'Tier 3' analyses.

There are two potential approaches to risk analysis. A 'bottom-up' approach, as widely used in risk identification, starts with the definition of challenging circumstances and then works backwards to seek to determine the likelihood of those challenging circumstances arising due to climate change. This is also known as the 'vulnerability approach'. More usually, however, risk analysis is 'top-down': it starts with the potential stressor – such as a specific change in climate – and follows this down through an organisation to determine impacts. This is also known as the 'hazard approach' to risk analysis, because it starts with the hazard. For reasons outlined below, this approach is usually easier to implement in practice than a bottom-up approach.

There are three key issues in estimating quantitatively the impacts of climate change for a specific risk:

- selection of the changes in climate that need to be considered;
- translation of these changes into relevant measures of impact;
- specification of changes in other relevant factors which might affect the impact.

There are two ways of defining changes in climate. The first is to undertake a *sensitivity analysis* to see how a system responds to different degrees of climate change. This is most feasible where the system depends on just one or two climate variables, such as sea level, temperature or precipitation. A sensitivity analysis could explore how the system responded if sea level were to rise by 10, 20, 30 or 40cm, or if temperature were to increase by 1, 2 or 3°C. A more complicated sensitivity analysis would look jointly at changes in temperature and precipitation (what happens if temperature goes up by 1, 2 or 3°C and rainfall changes by -10% or +10%). However, such an analysis is only really feasible if the system depends on at most two variables, and a sensitivity analysis is designed to see how sensitive a system is to change. It makes no judgement on whether the changes being considered are plausible. A sensitivity analysis can be undertaken as a preliminary risk analysis simply in order to see whether the system is sensitive to a change in climate.

The second approach is to use one or more *climate scenarios* and use a scenario planning approach. A scenario is a description of a plausible and internally consistent future climate regime. It is neither a prediction nor a forecast, and simply defines what could happen as a result of increased concentrations of greenhouse gases. Scenarios can be specified for just one variable (such as sea level rise), but scenarios typically define plausible changes in a number of climate variables through the year. Plausibility and internal consistency are key to a scenario. Plausibility is in many senses subjective, and is based on expert judgement and understanding of the types of change that are physically possible. For example, a scenario describing an increase in average temperature of 10°C would be regarded as physically implausible, as would one defining a sea level rise of 15m. Similarly, a scenario assuming no change in temperature in the future would also be regarded as implausible. Internal consistency means that changes in two or more variables must make sense together. Techniques used to construct plausible and internally consistent scenarios are discussed below.

Because a scenario is not a specific forecast, and because there may be several plausible future climate regimes at a place, it is important when assessing risk to use several scenarios. The number of scenarios which should be considered will depend on the level of detail required, resources available and the scenarios that are available. The Australian framework recommends using at most two climate scenarios in a risk assessment, but suggests using more (without specifying how many) if more detailed analysis is deemed necessary. The Californian framework refers directly to available Californian scenarios (up to four). As we shall see in more detail below, most scenario data sources designed for use by non-research organisations contain three or four scenarios. The range in estimated impacts of climate change can depend very much on the types and number of scenarios used.

Figure 5.8 illustrates the different indications of climate change risk when using different numbers of climate scenarios. It shows the estimated change by 2020 in a

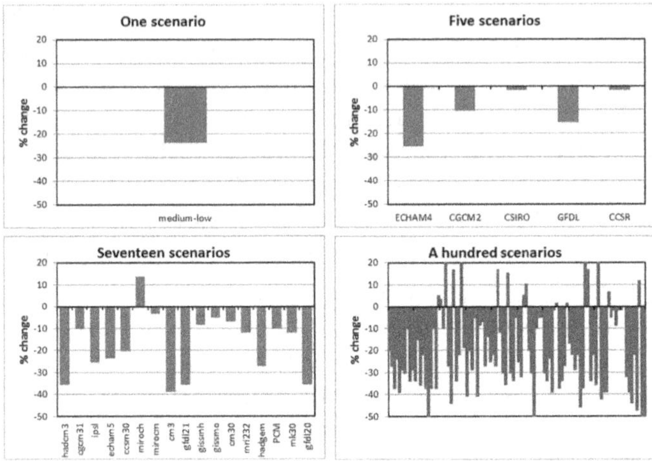

Figure 5.8 An example of impacts estimated with multiple scenarios. The graph shows change in a measure of low river flows by 2020 in a river in eastern England. The scenario labels represent the different climate models used to construct the climate scenarios

measure of river flow behaviour (relevant to farmers seeking to extract irrigation water). The top left panel shows the projected change if we just use one scenario – it shows that the flow measure would reduce by just over 20%. The top right panel shows another five scenarios; these show a range from minimal change to a reduction of around 25%. In comparison with these five scenarios, the first looks to be implying rather large impacts. The bottom left panel shows impacts under another set of 17 scenarios. Here, the greatest reduction is nearly 40% (suggesting that the first change is perhaps not so extreme), and one of the 17 produces an increase in the flow measure. The bottom right panel shows the changes in the flow measure across 100 plausible scenarios. The range in estimated impact clearly depends on which scenarios are considered.

If a sufficiently large number of scenarios are used then it is in principle possible to construct probability distributions of the likelihood of specific outcomes and undertake a *probabilistic risk assessment.* Figure 5.9 uses the same example indicator of impact as shown in Figure 5.8, but this time several thousand equally plausible climate scenarios have been used to estimate impacts. The estimated impacts have been ranked, and expressed as a probability distribution. This figure suggests that there is an approximately 55% chance that the flow measure will fall by more than 20%, a 35% chance that the measure will fall by more than 30%, and a 15% chance that it will fall by more than 40%. The more scenarios used, of course, the more complicated and time-consuming the analysis becomes. Moreover, likelihood

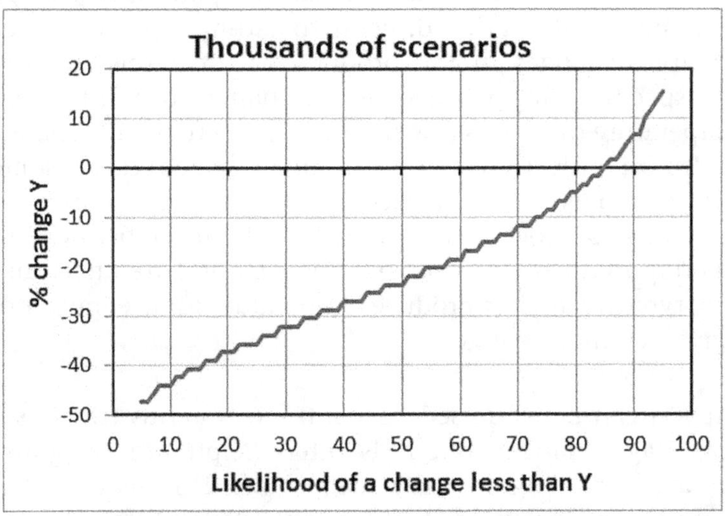

Figure 5.9 A probability distribution of climate change impact, based on thousands of equally plausible climate scenarios. The graph shows change in a measure of low river flows by 2020 in a river in eastern England, and can be compared with Figure 5.8

assessments are to be interpreted cautiously even when large numbers of scenarios are used. This is because there is no guarantee that the range of scenarios used represents the full range of plausible climates, or even represents the distribution of potential outcomes.

The second key issue in a risk analysis is the translation of the change in climate as specified by the climate scenario to a measure of impact relevant to the organisation. In some cases, there will be a direct relationship between a simple climate variable, such as temperature, into impact. For example, there may be a clear step change impact when temperature rises above some critical threshold. However, in the vast majority of cases, impact will be dependent on different climate variables or on the pattern of change over time, and may be difficult to estimate precisely. For example, changes in the productivity of a specific crop in a specific place will depend on changes in rainfall and temperature over the growing season, together with change in atmospheric CO_2 concentration. In such cases, some form of model will be needed to link climate to impact. A wide range of models exist – as will be discussed below – but it is important just to note here that models are uncertain; they typically do not produce exact values for the impact of a given climate change.

Impacts can be expressed in 'continuous' terms (such as a percentage change), but it is often helpful to categorise impacts according to their magnitude. This requires the definition of thresholds for different levels of impact, and the production of an 'impact rating matrix' or 'consequence table', as illustrated in Figure 5.10. Table 3.2 gives an example of an impact rating matrix for the UK (from the UK's Climate Change Risk Assessment).

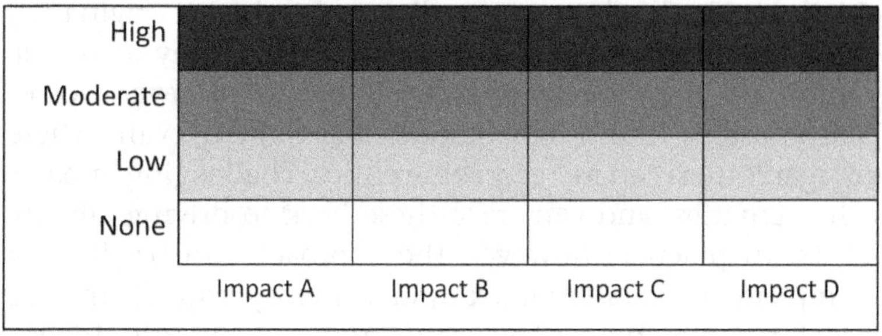

Figure 5.10 An impact rating matrix

The third key issue in a climate change risk analysis is to incorporate scenarios representing the other changes which might affect an organisation. These other changes include changes in population totals, changes in demand for a product or service, and regulatory changes which affect how an organisation operates. Many parts of the natural environment, such as ecosystems and hydrological systems, are also subject to pressures from land use change and pollution. It is important to include such changes where feasible for two main reasons. First, they may have a greater impact on the organisation than climate change; this would imply that dealing with them would be more important than coping with climate change. Second, they may affect how climate change impacts upon the organisation. For example, the effects of climate change may be greater if at the same time land use change has also led to a reduction in the amount of water in a river.

So far, this section has focused on the 'top-down' approach to estimating climate change risks. This starts with a set of climate scenarios, and works down through to their consequences. The 'bottom-up' approach starts by identifying critical changes in a system, and then works back up to see what sorts of climate change would produce these critical impacts. The approach is widely used in development studies (from where it originated) where the aim is to work out such things as why a specific community is prone to poverty, ill-health or severe losses from disasters. It also has the potential to be relevant where an organisation can identify some critical challenging impacts to its activities, and can trace these back to driving climate events. In practice, however, the approach can be difficult to implement. In a complex system it may require 'inverse modelling', which goes backwards from outcomes to drivers. Most models work the other way, and model how drivers lead to outcomes. Also, the approach still needs climate scenarios, because it has to see how plausible the identified climate changes are. Most quantitative risk analyses will therefore be undertaken using the top-down approach.

RISK EVALUATION

The final stage of risk assessment is to evaluate the 'significance' of risks identified during the risk analysis, in order to decide what to do next or to compare different risks. This stage allows the organisation to determine, for example, whether the effects of climate change on exposure to flood risk are worth worrying about, or whether the risks caused by increased flooding are more significant and therefore of a higher priority than those posed by higher temperatures.

If the impacts of climate change are 'small' (as defined by the impact rating matrix) under all the analysed scenarios for all potential risks, then risk evaluation is straightforward: climate change poses little risk. If the impacts are 'large' under all scenarios for each risk then the risks clearly need addressing, and the purpose of risk evaluation is two-fold: first, to determine the likelihood of these large impacts occurring, and second to prioritise each risk.

Risks are often evaluated and compared using a *risk matrix*. Figure 5.11 shows an example risk matrix, which categorises risk (for example due to increased flooding or higher temperatures) based on the magnitude of the consequences and the likelihood of the impact occurring. The categories for 'magnitude', 'likelihood' and 'risk', of course, have to be determined by the organisation undertaking the risk assessment. These will perhaps have been defined in the scoping or risk identification stages, and will depend on factors including coping capacity, critical thresholds and appetite for risk. Risk matrices are particularly good for comparing different risks.

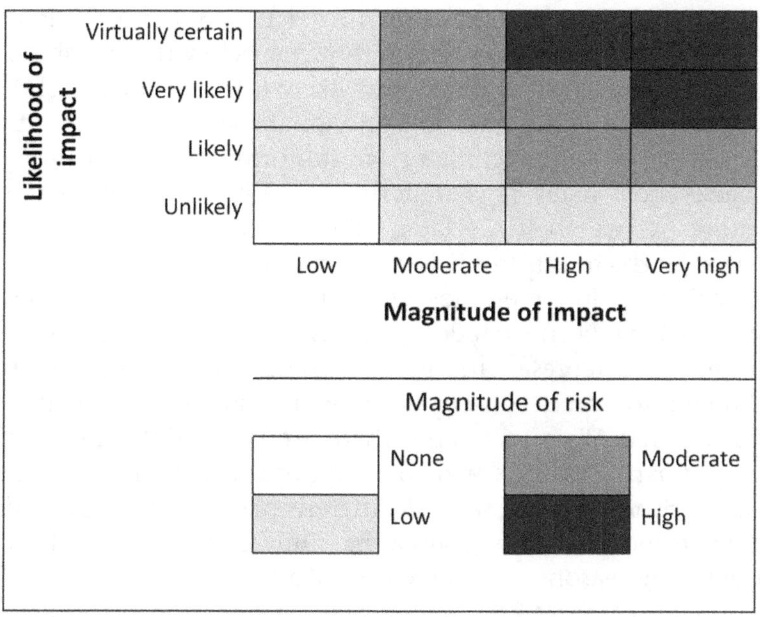

Figure 5.11 An example of a risk matrix, showing the relationship between magnitude of risk and the likelihood and magnitude of impact

In the context of climate change, a risk matrix is most useful where impacts are caused by discrete events and where the impacts do not vary with the magnitude of the event. Such events could include major equipment failure, loss of a supplier or indeed an extreme climate-related event such as a flood. Climate change could mean that some discrete significant event occurs either for the first time or more (or less) frequently. The risk matrix would be used to compare the impacts under such an event (for example in terms of machine down time or lost production) with the chance of the event occurring (or occurring more frequently) in the future.

Suppose that a critical threshold for an organisation is passed if the river flow measure shown in Figures 5.8 and 5.9 falls by 30%, and it does not matter how far below the threshold it falls. We need to estimate how likely it is that the measure will fall below this threshold. A climate scenario represents a 'plausible' future climate, but unfortunately we cannot specify how likely it is that this climate will occur in the future. As we shall see later, climate scenarios are usually based on the results from climate simulation models forced by specific emissions scenarios. Once any obviously 'bad' climate models have been excluded (such as those that produce very implausible representations of current climate), it is very difficult to reach consensus on which models are 'better' than others. This is because there are many different ways of comparing models with observations of real climate, the comparisons can be made at different places and scales, and a given model may be simulating climate at a place well for the wrong reasons. It is therefore usually assumed in practice (though not necessarily in research) that each 'good' model, and therefore the scenario produced from it, is as likely as any others. Sources of climate scenario data usually state explicitly that each scenario provided is as likely as any of the others.

We can, however, make some assessment of the relative likelihoods of specific climate *outcomes or impacts* if a number of scenarios are used. We can simply estimate the likelihood of the flow measure falling by more than 30% by counting the number of scenarios considered where this happens. Likelihood can then be plotted against magnitude on Figure 5.11 to evaluate risk for this particular dimension of impact. The estimated likelihood of the event occurring, however, depends on the set of scenarios considered. If we just consider six scenarios (the top two panels in Figure 5.8), then the chance of the river flow measure falling by more than 30% is effectively zero. If we use the wider set of 17 scenarios (the bottom left panel) the chance is around 20%, and if we use a very large number of scenarios (Figure 5.9) the chance is approximately 35%. We can take this uncertainty into account when plotting on the risk matrix. In practice, however, we often will only have a small number of scenarios so cannot quantify this uncertainty effect, but we should be aware that our estimate of likelihood should not to be taken too literally. In principle we can also include uncertainty in the magnitude of our estimate of impact when plotting on a risk matrix.

More commonly, however, the impacts of climate change depend on the magnitude of the change, and likelihood therefore varies with magnitude. In these circumstances it is therefore not possible to use a risk matrix to come up with some single evaluation of risk. Instead, it is more appropriate to construct some form of 'risk profile' for each risk, showing magnitude of impact against some indication of likelihood. In practice, it is not possible to estimate the likelihood of a given exact magnitude of impact, so instead it is necessary to seek to assess the likelihood of an impact either being within defined ranges or beyond defined thresholds.

Turning again to the example in Figures 5.8 and 5.9, it is not appropriate to ask for the likelihood that the flow measure will reduce by exactly 30%, or exactly 15%. Instead, it is necessary to ask either for the likelihood that the flow measure reduces by, say, between 15 and 30%, or by between 30% and 45%, and so on. This is illustrated in Figure 5.12, which shows the approximate likelihood that impacts will be within specific ranges. Again, three different sets of scenarios are shown to illustrate the sensitivity of estimated likelihoods to the amount of information used; in practice it is likely that only one set of scenarios will be used. In this example, there is a 25-30% chance that the flow measure will reduce by between 15 and 30% (when at least 17 scenarios are used), and a 25-35% chance that the measure will reduce by between 30 and 45%. If only six scenarios are used, then there is an approximately 50:50 chance that the reduction will be 15-30% or 30-45%. The estimated likelihoods are therefore typically highly dependent

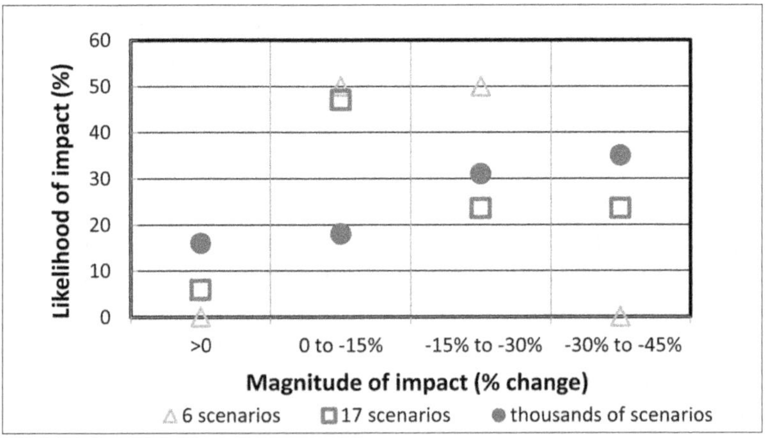

Figure 5.12 **Example risk profiles, for the impact indicator shown in Figures 5.8 and 5.9. The profiles show the relative likelihood of different categories of magnitude of impact, using three different sets of climate scenarios**

on which, and how many, scenarios are used. This uncertainty is important when evaluating the risks of climate change.

INFORMATION FOR RISK ASSESSMENT

The previous section has outlined the stages in a climate change risk assessment, culminating in an assessment of the relative likelihoods of different amounts of impact. Climate scenarios are clearly extremely important in a risk assessment, even if the assessment is primarily based on a determination of the sensitivity of a system to a change in climate. Scenarios provide some boundaries on the extent of climate change that should be anticipated. 'Bottom-up' vulnerability approaches, which start from the identification of challenging circumstances and work backwards to see if climate change would cause problems, need some information on potential future climates.

This section looks in more detail at climate scenarios, how they are constructed, and their limitations. Different assessments need different types of scenarios with different degrees of detail and perhaps different relevant climate variables. The section also considers currently available sources of climate scenarios. A climate change risk assessment also, of course, needs to use some tools to translate changes in climate to relevant measures of impact. The section therefore concludes by looking at the key issues involved in estimating impacts under a given climate scenario.

CLIMATE SCENARIOS

A climate scenario is a description of a plausible future climate. It is not a forecast of what the future climate will be like, or a prediction of future conditions; it is a description of what the future *could* be like. A climate scenario is typically contingent on some assumptions, most usually an assumed scenario of future growth in greenhouse gas and other emissions. In the broadest sense, a climate scenario would include descriptions of all the relevant dimensions of climate change. These could include not only weather variables, but also the concentration of greenhouse gases in the atmosphere. Marine climate scenarios define sea level rise, and may also define future ocean acidity as well as ocean temperature.

There are two broad approaches to constructing plausible climate scenarios. The 'analogue-based' approach assumes that the future climate at a place will be similar to the climate experienced either at that place sometime in the past, or experienced currently at another place. The 'model-based' approach uses a suite of models driven by emissions scenarios to project plausible future climates.

A 'temporal analogue' is a period from the past which is assumed to have the properties of the future climate at a place, while a 'spatial analogue' is provided by another place. Whilst such analogues can be intuitively appealing and can help in the visualisation of future climate change, there are two major problems with them. First, there needs to be some logic in their selection as indicative of future climate, otherwise analogues become uninformative sensitivity studies. This logic could be based on physical principles, or perhaps on the results of model projections of future climate. For example, it has frequently been assumed that in the future the climate of southern England will be much more like that of the southern France now; warmer throughout the year, with less rain during summer. There are physical reasons for believing this to be plausible, because the anticyclonic conditions currently pertaining over the Mediterranean in summer could extend further northward and push rain-bearing depressions to the north of Scotland. In contrast, it is unlikely that southern England will ever experience a similar climate to south eastern Europe which is also currently warmer than England, simply because England will always be on the wet westerly edge of a continent. The second major problem with analogues is that they will always be partial, and that other important differences between times and places may be more significant than the differences in climate. Whilst the climate of southern France might be a reasonable analogue for future climate in southern England, it would be a step too far to conclude that the future landscape of southern England would look like that of southern France (because the geology would remain different), and several steps too far to suggest that the lifestyles of the English would become like those of the French. Whilst analogues can give some indication of what the future might look like, they therefore have some significant limitations. Perhaps the most useful analogue is provided by past extreme events at a place. These can be used to 'stress-test' the sensitivity of current systems and operations to extreme events in order to understand potential vulnerabilities.

The vast majority of climate change risk assessments are based on climate scenarios which are derived from model projections of future climate. As introduced in Chapter 2, these projections involve two stages. The first is the definition of some assumed amount of external forcing on the climate system – through scenarios for greenhouse gas emissions – and the second uses a computer simulation model to estimate the climatic consequences of this forcing. The simulated climates are therefore conditional on the assumed amount of forcing.

Between 2000 and 2010 most projections of potential future climates were based on the four IPCC SRES scenarios.[14] These made different assumptions about future population, economic development and the use of energy sources. Emissions were greatest under the A1FI ('Fossil-intensive') emissions scenario, and lowest under B1. All the emissions scenarios were assumed to be equally plausible, although in practice it would be unlikely that emissions would follow the lowest scenario unless there were significant changes in the way energy is used. Since 2010 most climate model projections have been based on four Representative Concentration Pathways (RCPs[15]), which were designed to characterise different patterns of future forcing on the climate system. They are not assumed to be equally plausible, and indeed the lowest pathway requires very considerable efforts to reduce future emissions.[16] Emissions scenarios and RCPs define the assumed future concentration of greenhouse gases.

Global climate models are used to simulate the evolution of climate under a given emissions or concentration pathway. As outlined in Chapter 2, these models are numerical representations of the physical equations describing the movement of mass and energy around the globe in the atmosphere and oceans. The models are the same in principle as those used for short-term weather forecasting, but unlike the forecast models are not continually updated and 'nudged' with observed data. There are currently around 20 to 30 global climate models which are used to make projections of future

14 IPCC (2000), *Special Report on Emissions Scenarios*. Intergovernmental Panel on Climate Change.

15 See Chapter 2.

16 IPCC (2014), Summary for Policymakers. In: *Climate Change 2014: Mitigation of Climate Change. Contribution of Working Group III to the Fifth Assessment Report of the Intergovernmental Panel on Climate Change*. Edenhofer, O., et al., eds. Cambridge University Press.

climate, although because these share many components the models are not all independent. The very broad patterns of change in climate are consistent between the models, but there are important regional differences. All models, for example, project greater warming over land than over sea, but the rate of warming for a given amount of forcing differs between models. All models project that wet regions tend to get wetter and dry regions drier, but the amount of change varies between the models and the precise location of wet and dry regions differs. For a given emissions scenario, there is therefore a range of projections of change in global climate.

Global climate models are, as their name suggests, designed to simulate the broad pattern of global circulation and weather. They operate on a grid resolution, and although this resolution has improved over the years global models typically currently operate on scales of the order of several hundreds of kilometres. They therefore do not endeavour to simulate accurately regional and local climate. These may be very strongly influenced by local topography, or the configuration of land and sea, or the operation of locally significant, but globally unimportant, processes (such as evaporation from local lakes). It is therefore not appropriate to estimate future climate at a place simply by extracting projections from a global climate model.

A great many 'downscaling' techniques have been developed to try to extract locally plausible information from global climate models. They fall into three groups. The first uses a much higher-resolution model 'nested' within the global model to add local detail; the simulations from the high-resolution model are then used to represent current ('baseline') and future climate. The second uses a range of statistical techniques to interpolate down to finer spatial scales or correct for bias in simulated local climate. Both of these groups of techniques are complicated and require considerable computing resources. The third group of techniques is much simpler. These basically assume that even if the absolute climate at a place is poorly simulated by a global climate model, the simulated *change* in climate (relative to a baseline) at a place due to increased emissions is plausible. These changes can then be added to observed local climate data for that same baseline period to construct data representative of future conditions. Climate scenarios designed for practical applications tend to be based on this third approach, whilst scenarios used in research studies often use one of the other two approaches. For a given climate model, different downscaling techniques to construct local-scale scenarios will potentially give different results. However, for practical applications in risk assessments this uncertainty does not matter. There are two reasons for this. First, climate scenarios are descriptions of plausible future climates and not predictions: individual scenarios should not be taken too literally. Second, risk assessments should use two or more climate scenarios, constructed from different climate models. The difference between climate models is usually (but not necessarily always) greater than the difference between downscaling techniques.

Climate models simulate all aspects of climate at the resolution at which they operate. However, the plausibility of these simulations differs between variables. Models are generally good at simulating average monthly temperatures across a region, but less good at simulating the variation in temperature from day to day at the local scale. Scenarios for change in monthly mean temperature are therefore more credible than scenarios for changes in day-to-day variability or periods of extreme high or low temperatures. Climate models are less good at simulating regional precipitation than regional temperature, partly because precipitation is strongly controlled by atmospheric circulation patterns which can be difficult to simulate, and partly because precipitation may be influenced by local processes such as convection which operate at a much finer scale than the models. Models are generally better at simulating average climate than variability from year-to-year, so scenarios of changes in year-to-year climatic variability are less robust than scenarios for change in average climate. Whilst current climate models simulate well the broad features of weather, the vast majority of global models are not capable of simulating individual storm events because these are typically generated at scales finer than the models operate, so cannot be used to construct scenarios for changes in the frequency or intensity of storms (including hurricanes and typhoons).[17] Most climate scenarios designed for practical application therefore only define changes in mean monthly temperature and mean monthly precipitation.

Different climate models forced with the same emissions or concentrations will produce different climate scenarios, especially for precipitation. It is, however, extremely difficult to determine the relative plausibility of different climate scenarios. The skill of weather forecast models can be quantified because observations soon become available to compare with forecasts, but of course we do not have observations yet from the future; climate projections can therefore not be validated. Some indications of the plausibility of projected future climates can be inferred by seeing how well models simulate current climate and its variability. However, there are many possible ways of comparing simulations and observations (which weather variables, and over what scales?). Even if a model simulates current climate well, it may not necessarily

17 A few very high-resolution global models are currently (2014) capable of simulating individual storm events.

simulate future climate well; it could be getting the current climate right for the wrong reasons. It is possible to identify – and eliminate – models that have particularly strong biases by comparing simulations with observations, but in general it is not possible to objectively weight different models on the basis of their performance in simulating current climate. It is therefore typically assumed that all climate scenarios (for a given emissions scenario) are equally plausible.

Scenarios for sea level rise are constructed from climate models, supplemented by projections made by ice sheet models. Most sea level rise occurs because of the expansion of sea water as it gets warmer, and this is simulated explicitly in most global climate models. The other major contribution is from the melting of ice in glaciers and the ice sheets of Greenland and Antarctica. Climate models do not include these processes explicitly, because they are not directly relevant to the simulation of weather and climate. Different models are therefore used to simulate changes to the mass balance of glaciers and ice sheets, and therefore their net contribution to sea level rise. As is the case with climate models, there are several different ice sheet models and so there is a range of potential sea level rises for a given emissions scenario.

Model-based climate scenarios therefore vary with the forcing emissions or concentration scenario, and the climate model used to translate forcing into climate. The climate system responds relatively slowly to forcing, so in practice there is little difference in the amount of climate change over the next 30 or 40 years between different emissions scenarios (see Figure 2.5). The difference between climate models is considerably greater than the difference between different emissions assumptions. From around the 2060s, however, there is a more clear difference between emissions assumptions, so risk assessments over this time horizon need to be more explicit about assumed future emissions.

CLIMATE SCENARIO SOURCES

As mentioned above, there are currently 20 to 30 global climate models which have been used to project future climates, and many different techniques and models have been used to create scenarios for estimating impacts at the local and regional scale. The vast majority of these scenarios have been constructed for research purposes, and have not been designed for use outside the research environment. They can be very difficult to use, and require considerable meteorological and technical knowledge on the part of the user. There have, however, been several attempts to produce climate scenarios for use in practical risk assessments.

Data from global climate models are publicly available from the IPCC's Data Distribution Centre,[18] but this is designed primarily for climate change researchers and the data are provided in a complex data format. The full report of IPCC AR5 Working Group I includes an annex with an atlas showing projected changes in regional December-February and June-August temperature and precipitation. Global temperature and precipitation scenarios at a spatial resolution of 2×2° can be downloaded from the World Bank climate portal,[19] from nine climate models under two of the SRES emissions scenarios. This website is aimed at development practitioners and policymakers.

Several countries have produced climate scenarios for use in impact and risk assessments, and these are summarised in

18 www.ipcc-data.org.
19 sdwebx.worldbank.org/climateportal/index.cfm?page=climate_data.

Table 5.2. The formats and types of scenarios provided vary considerably, and the baseline periods differ. The Finnish scenarios represent low, medium and high emissions, and each emissions scenario uses a different climate model. The Californian website presents projections from four different climate models, plus the mean across all models, for each of three different rates of temperature change corresponding to low, medium and high emissions. The Australian scenarios (*Ozclim*) are provided at two levels. The three 'entry-level' scenarios are based on three different climate models scaled to low, medium and high emissions, whilst 'advanced-level' scenarios can be constructed from a wider range of climate models and include a wider range of climate variables. The American national website provides scenarios by region from a wide range of climate models, but does not explicitly provide guidance on which models to use. The Canadian website is similar, providing many different climate models but little guidance to help users. All these national scenarios provide information for a range of different future time horizons.

The Dutch scenarios (*KMNI'06*) are different in principle to the scenarios provided by the other countries, in that they are not based explicitly on output from specific climate models. Instead, they were constructed using expert judgement, and differ by the assumed change in temperature (+1°C or +2°C by 2050, relative to 1990), and whether or not circulation patterns change to produce wetter winters and drier summers.

The UKCP09 scenarios for the UK are also significantly different in concept to scenarios produced in other countries. Previous UK climate scenarios (produced in 1998 and 2002) were, like other national scenarios, based

Table 5.2 National climate scenarios

Country	Website	Summary	Variables
UK *UKCP09*	ukclimateprojections.metoffice.gov.uk/	Probabilistic climate projections	Multiple climate variables, plus marine variables and sea level rise
Finland	ilmasto-opas.fi/en/	Three scenarios, representing low, medium and high emissions	Temperature and precipitation
Netherlands *KMNI'06*	www.knmi.nl/climatescenarios	Four scenarios, winter and summer	Temperature, precipitation, windspeed and sea level
Australia *Ozclim*	www.csiro.au/ozclim/	Entry level: three scenarios based on three climate models and three emissions scenarios. Advanced level: Twenty three climate models	Entry level: temperature and precipitation Advanced level: multiple climate and marine variables
Canada	www.cccsn.ec.gc.ca/	Many climate models	Multiple climate variables
USA	scenarios.globalchange.gov/	Multiple scenarios, from several climate models	Multiple climate variables, plus sea level rise
California	cal-adapt.org/	Four climate models, plus mean, and three rates of temperature change	Temperature, precipitation, snowpack, wildfire risk and sea level rise

on individual climate model simulations scaled to different assumptions about future emissions. The UKCP09 scenarios, however, are probabilistic. They are provided in three basic forms. The first is a suite of probability distributions for changes in individual climate variables (such as temperature and precipitation) and for some pairs of climate variables, for each of three emissions scenarios. These can be used to give an indication of the potential range in change in climate at a point. The second is a set of thousands of equally plausible scenarios (again for each of three emissions scenarios) providing, at a point, consistent changes in a large number of climate variables. These scenarios can be used with an impact model to produce probability distributions of the impacts of climate change – and indeed these were used to produce Figure 5.9. This is a very powerful tool, but the probability distributions need to be interpreted carefully (as outlined above) and can be time-consuming to apply. Users can use a subset of the scenarios, but of course different subsets will produce different estimates of the distribution of impacts. The probability distributions, and the sets of thousands of scenarios, were produced in a complicated process involving multiple climate models and statistical emulation techniques in order to sample as wide a range of uncertainty as possible. The third set of scenarios are based on eleven of the climate models used to construct the probabilistic scenario sets; these provide more comprehensive and spatially coherent scenarios, but require much more expertise from the user. The UKCP09 scenarios also include projections for change in sea level around the UK coast. These are not probabilistic, and represent low, medium and high changes; a 'high++' scenario represents a plausible, but unlikely, upper bound. From each set, scenarios can be constructed for a sequence of different future 30-year time periods, starting from 2010-2039.

TRANSLATING CLIMATE CHANGE INTO IMPACTS

In some cases it may be possible to infer the impacts of climate change directly from the climate variables included in climate scenarios, but more frequently it will be necessary to use some form of model to translate changes in climate into impact. This may be because the impact depends on the combined effect of several climate variables, or it depends on how climate changes through the year. The impacts shown in Figures 5.8 and 5.9 were estimated using a hydrological model run with different climate scenarios.

Impact models are not perfect, and in many instances the translation of climate change to impact may be very uncertain. The credibility of estimated impacts therefore depends not only on the plausibility of climate scenarios, but also on the plausibility of the models used to translate climate to impact. In principle this can be quantified by using several different impact models – in the same way that climate scenario uncertainty can be represented by using scenarios from different climate models – but in practice this is rarely done. Nevertheless, impact model uncertainty is another reason why estimates of the impact of climate change should not be taken too literally.

In many cases, climate will not be the only thing that is changing, and where other drivers are important it is necessary to include them in the assessment of impacts. This may require the development of additional scenarios for possible changes in these non-climatic drivers, which then need to be mixed and matched with the climate scenarios. Whilst this adds to the plausibility of the projected impacts of climate change – and therefore any risk assessment – it can add substantially to the assessment burden.

CONCLUSION

This chapter has outlined the stages involved in a climate change risk assessment. In practice an assessment may not follow these steps explicitly, and sometimes some of the steps may be taken simultaneously. However, the chapter has demonstrated a number of points.

- It is important to plan a risk assessment robustly, and to follow procedures that are consistent with other risk management processes used in an organisation. This makes it easier to compare the risks (and opportunities) posed by climate change with the other challenges facing an organisation.
- There are several different ways of assessing the impacts of climate change on an organisation, with a clear distinction between a 'bottom-up' approach which seeks to identify vulnerabilities, and a 'top-down' approach which follows a change through an organisation to identify consequences. The 'top-down' approach explicitly requires the use of climate scenarios, and the selection and definition of these is critical to a risk assessment. The 'bottom-up' approach also needs to use climate scenarios in order to assess the significance of any identified vulnerabilities.
- Risks may come through the effects of climate change on an organisation, its supply chain and its customers, or may come indirectly through the actions of regulators and others in response to climate change.
- Although climate scenarios are crucial to a risk assessment, it is important not to take them, or the estimated impacts, too literally. Scenarios are simply plausible descriptions of

future climates, and not predictions. Different scenarios can give quite different indications of potential risks, and different sets of scenarios can give different estimates of the range and likelihoods of potential impacts.

This cautious scepticism is important when deciding how to deal with any risks posed by climate change – which is the subject of the next chapter.

6 Coping with the Risks of Climate Change

INTRODUCTION

There are two main stages in risk management. The first – identifying and evaluating risks through risk assessment – is discussed in Chapter 5. This chapter concentrates on the second stage: coping with the risks. This second stage is often termed 'risk treatment' or 'risk control'. Coping with risks posed by climate change is generally known as 'adaptation'. Sometimes the phrase 'risk mitigation' is used in risk management to refer to the treatment of risk, but in the climate change context this can be confusing because the term 'mitigation' here means reducing emissions of greenhouse gases to slow the rate of climate change, rather than coping with change that does occur.

In principle, organisations deal with risk and adapt to changing circumstances all the time. Climate change is different in three main ways. First, the consequences have not necessarily been experienced yet: they occur – or might occur – some time in the future. Even if the decision to think about adaptation is triggered by the occurrence of some extreme weather event, that decision is strongly influenced by the extent to which it is believed that conditions will change in the future. Organisations may also find it difficult to learn from their experiences about what works and what does not, because the impacts being managed are in the future. Second, the consequences of climate change are uncertain, particularly at the scale at which an organisation operates. Third, as more information is acquired over time, estimates of what may happen in the future, and therefore what needs to be done, can be revised and updated.

This chapter covers four issues:

- Approaches to adaptation, given that the future is uncertain;
- The identification of potential options for adaptation;
- The design and appraisal of adaptation options;
- Monitoring and evaluation.

The risk assessment and management frameworks shown in Figure 5.3 all implicitly or explicitly separate out option identification and option evaluation/appraisal as distinct stages in risk management. Chapter 3 identified a number of barriers to adaptation (Figure 3.8), which have the potential to significantly constrain the adaptation options that are identified and their evaluation. A key part of coping with climate change is recognising and seeking to overcome these barriers.

APPROACHES TO ADAPTATION

There are two main ways of looking at adaptation to climate change. The first is to see climate change as a specific challenge, and to determine how to adapt to that challenge: to ask, for example, how we can continue to provide the same service under a changed climate. The second is to see climate change as one factor that needs to be included in a review or investment decision. Organisations are continually adjusting to altered circumstances and planning new developments, and climate change can be seen as simply one part of this. The question an organisation might ask could therefore be 'how can we account for climate change risks in our investment planning'.

Regardless of whether adaptation to climate change is seen as a specific task or as part of a larger review or planning process, there are several different ways of approaching adaptation. These affect the design of adaptation options, and how adaptation options are evaluated. There are five basic approaches:

- Cope with climate change by adapting better to current vulnerabilities;
- 'Predict and plan': predict the future and plan accordingly;
- Take a precautionary approach: plan for the worst;
- Follow an adaptive approach: hedge and review;
- Implement win-win measures.

All these approaches have strengths and weaknesses.

ADAPT TO CURRENT VULNERABILITIES

The risk assessment process is likely to have identified a set of vulnerabilities to current climatic variability and extreme events. An organisation may have discovered that its operations would be significantly affected by a flood, or that its supply chain may be disrupted by a drought affecting crop production, or that a heatwave may adversely affect worker productivity. Even with no change in climate, measures to reduce exposure or vulnerability to flood, drought or heatwave would therefore reduce risks to the organisation. Adapting better to current variability in weather – improving resilience – therefore goes part of the way to adapting to future climate change. However, it is not necessarily sufficient, for two reasons. First, future extreme events may be different in character to events that occur at present. They may be more intense or more persistent, or occur closer together; they could even become *less* frequent. Second, what seems like an appropriate response to current vulnerability may not be an appropriate response to future vulnerabilities. Suppose for example that a facility is exposed to flood risk at present. One response could be to build a protecting wall around the facility to keep flood water out, and that could encourage further investment in the facility. However, as climate changes the chance of flood water breaching the defences could increase, or an increased frequency of heavy rainfall events could mean more water is ponded within the facility, and because investment in the facility has increased losses would be higher. In this case, the adaptation to current vulnerability would have led to increased exposure to loss in the future. Another adaptation – such as making the facility itself less vulnerable to being disrupted by water – might not be so appealing in the short term, but might be better in the long term.

PREDICT AND PLAN

A second potential approach is to make a forecast of the future effects of climate change, and design an adaptation plan accordingly. This may include incorporating some form of 'safety factor'. Such an approach has traditionally been widely used in planning over a range of time-scales, and there is accordingly a large literature and much accumulated experience. However, whilst the approach might be appropriate over the weekly and perhaps monthly scales of weather forecasts, there are substantial problems in applying the approach over the medium and long terms. First, we cannot make forecasts of future climate – as explained in more detail in Chapters 2 and 5 – and can only make plausible projections conditional on specific assumptions. Second, there can be considerable uncertainty in these projections, and no individual projection is necessarily more plausible than another. It would therefore be rather risky to plan on the assumption that any one individual climate scenario turns out to be 'true'. If the future were to turn out to be different, then the adaptation made may be quite inappropriate; it could be completely ineffective, or vastly over-engineered and expensive. 'Predict and plan' encourages us to put all our eggs in one basket. It does not encourage flexible or adaptable management.

Chapter 5 introduced probabilistic approaches to risk assessment, which involve the construction of probability distributions of potential impact from the use of many climate scenarios. Whilst these provide very useful information on the distribution of potential impacts, there can be a strong temptation to use a probabilistic approach to 'predict and plan'. For example, it may appear tempting to plan to be able to cope with the impact as represented by the 90% value on the probability distribution. In the example given in Figure 5.9 there is a 10% chance that the impact indicator will reduce by more than 45%, so there is a 90% chance that the impact will be less than a reduction of 45%. On first sight, it could make sense to plan for a reduction in flows of 45% and be 90% confident that this will be sufficient. However, our probability distribution is itself uncertain (we do not know if there really is a 90% chance that the flow measure will reduce by less than 45%) and, as noted above, focusing our planning efforts on one number may mean we do not consider flexible or adaptable approaches.

PRECAUTIONARY APPROACH

The precautionary principle has been widely defined, and basically implies that action should be taken to reduce potential adverse consequences even if those consequences have not been clearly defined or quantified. A precautionary approach therefore in a sense seeks to 'plan for the worst', and can be seen as a variant on the traditional 'predict and plan' approach: it simply uses the 'worst case' scenario. This approach may be appropriate where the risks are extremely significant, or where we know what the 'worst case' might be, but like 'predict and plan' may not necessarily be appropriate in the context of climate change. Whilst we can estimate impacts under the most extreme scenario (for example with the highest temperature increase or largest reduction in rainfall), we cannot say whether these really represent the worst case. The most extreme scenario available may be highly unlikely or implausible, or on the other hand may not actually represent the most extreme change that could plausibly occur.

ADAPTIVE MANAGEMENT

Adaptive management is defined as the sequential and continual process of making the best decision at each decision point,[1] and is an important potential strategy for handling uncertainties. It can be seen as a strategy of 'hedge and adjust' rather than 'predict and plan'. Decisions are reviewed as more information becomes available, and plans and actions adjusted. Central to adaptive management is the idea that it is important to keep all potential future options open, and not to take decisions that might constrain future options. Adaptive management involves testing potential adaptation options against a range of scenarios.

Adaptive management is particularly suited to circumstances where decision lead times are short, where actions can be easily updated or revised, or where actions can be deferred. It is more difficult to apply where lead times are long and where, once committed, decisions are difficult to change. This may be the case with capital investments with long payback times or which are difficult to upgrade. For example, once installed an air-conditioning unit may be expensive to upgrade, or a building once completed might be expensive to refit.

WIN-WINS

Climate change is only one of a number of challenges facing an organisation. Some actions to cope with climate change risks may affect these other risks, and indeed some measures to address other risks may affect vulnerability to climate change. 'Win-win' measures are those which have multiple benefits. For example, office buildings can be designed to

1 Willows, R. and Connell, R. (eds) (2003), *Climate Adaptation: Risk, Uncertainty and Decision-Making*. UKCIP, Oxford.

be less sensitive to heatwaves (adapting to climate change), whilst using less energy (reducing costs) and having better indoor air quality (improving worker productivity). Reducing water use lessens vulnerability to water shortages, and reduces costs. Win-wins are best identified by considering climate change alongside other risks and opportunities.

IDENTIFYING ADAPTATION OPTIONS

The previous section introduced some ways of thinking about adaptation to climate change, given that impacts will be experienced in the future and that future is uncertain. This section focuses on the adaptation options that are available to an organisation. The precise mix will of course depend on the nature of the organisation and the nature of the risks and opportunities to which it is exposed, including the other non-climatic drivers of change. However, it is possible to make some general comments. First, a distinction is frequently drawn between measures which seek to enhance *adaptive capacity*, and specific *adaptation actions*.

ENHANCING ADAPTIVE CAPACITY

Adaptive capacity represents the ability of an organisation to adapt to climate change. It is impossible to measure, but possible to conceptualise and perhaps represent by indicators. The adaptive capacity of an organisation is determined by both internal and external factors (Table 6.1). A key internal factor is the access of the organisation to information on the potential implications of climate change, and the interpretation within the organisation of that information. In organisational theory terms, this is the ability to recognise and interpret stimuli to support learning. Related to this are

Table 6.1 Factors influencing the adaptive capacity of organisations

Internal	**External**
Access to information: ability to detect and interpret signals	Regulatory context
Skills and knowledge	Market context
Organisational culture and routines	Availability of support and guidance
Access to resources	Licence to operate

the skills and knowledge of impacts and adaptation options available to the organisation, both internally and through support networks and consultants. Organisational culture and routines define how and why the organisation works. It includes such dimensions as corporate attitudes towards risk, the prevalence of fixed and established routines for operating and making decisions, and also the extent to which the organisation follows specific risk management procedures. Access to resources includes the ability to raise funds within the organisation and from outside sources.

The capacity of an organisation to adapt, however, is strongly influenced by the context in which it operates. Four dimensions of this external context are shown in Table 6.1. The regulatory context may constrain what an organisation can do, but could – through reporting requirements for example – stimulate action. The market context may significantly constrain adaptation if consumers are extremely price sensitive or competitors are less exposed to climate change risks. The availability of support and guidance from external organisations can open up opportunities to an organisation, but the lack of support and guidance may hinder effective adaptation. Finally, an organisation's 'licence to operate' may influence its scope for adaptation. Customers of an organisation may expect a specific standard of service to be maintained, or expect an organisation to operate in a specific way (such as following specific ethical principles). This may allow organisations to innovate, but may constrain adaptation options.

In principle, an organisation has the scope to affect the internal factors shown in Table 6.1 in order to enhance its adaptive capacity. It is much more difficult for an organisation to affect the external factors. It might be able to influence its licence to operate through shrewd marketing. The regulatory context and availability of support and guidance can potentially be influenced by lobbying on the behalf of individual organisations or industry groups.

ADAPTATION ACTIONS

Table 6.2 presents one classification of specific adaptation measures which might be available to an organisation,[2] either as stand-alone adaptations to climate change or as components of a more general investment strategy and plan.

The first option – 'bear the risk' – is a conscious decision to do nothing. Under this option, the organisation accepts that impacts may occur in the future, and aims either to accept them when they arise or seek to 'muddle through' using established procedures. 'Avoiding the risk' means seeking to move away from a hazardous location, for example, or moving out of a climate-sensitive business. 'Spreading the risk'

2 It is modified from a widely-used (and frequently modified) classification presented by Burton, I. (1996), The growth of adaptation capacity: practice and policy. In Smith, J. et al. (eds), *Adapting to Climate Change: An International Perspective*. Springer-Verlag, New York. 55-67.

Table 6.2 A classification of potential adaptation options

Type of option	Examples
Bear the risk	Accept the potential for impact, and aim to absorb the consequences or 'muddle through'
Avoid the risk	Move away from a hazardous location Move out of a business area
Spread the risk	Purchase insurance against loss Diversify supply sources or production sites
Lessen the risk by preventing the impacts	Install structural protection scheme
Lessen the risk by reducing exposure	Design 'hazard-proof' facilities Implement a warning and emergency plan
Learn about the risk	Undertake research into risks and adaptation options
Exploit the opportunity	Take advantage of the positive benefits of climate change Implement win-win measures with multiple co-benefits

involves buying insurance against possible loss (spreading the loss for an organisation over time), and also diversification. An organisation may be able to introduce additional sources of supply into the supply chain or develop additional production or operational sites so that an event or change in climate is less disruptive.

The next two categories of option in Table 6.2 – 'lessen the risk' – have similar outcomes, but achieve them in different ways. The first category ('lessen the risk by preventing the events') comprises measures which seek to reduce the magnitude or frequency of damaging events. These would include measures such as flood defences to lower flood water levels or keep water away from property, or water supply schemes to maintain supplies during droughts. Such measures are typically provided by public authorities, although property-scale measures (for example on-farm irrigation) can be implemented by individual organisations. The second category ('lessen the risk by reducing exposure') consists of measures that do not try to manage the events or counteract the effects of climate change, but instead seek to reduce impacts when events or climate changes occur. These measures include developing 'hazard-proof' buildings which do not suffer damage when an event occurs, changing crop types or varieties to cope with altered climatic conditions, and the adoption of warning systems and emergency plans. Such measures are typically adopted by individual organisations, but important public measures in this category are spatial land use planning (for example discouraging development in flood-prone locations) and the provision of warning services. Public agencies also have a role in setting standards, for example for buildings and emergency plans.

The next category of options in Table 6.2 includes measures to understand more about risk and potential adaptation measures. In one sense this can be seen simply as another loop in the risk assessment process – moving from a high level to a more detailed assessment – but it is more appropriately regarded as an adaptation option because it requires additional time and effort. The risk assessment process is effectively put on hold whilst the research and learning takes place.

The final category of options in Table 6.2 highlights measures which are designed to exploit the opportunities posed by climate change. These include measures to take advantage of the positive effects of climate change for an organisation (such as an increased demand for its products, reduced operating costs or the availability of new sources of supply), as well as measures which have multiple co-benefits. For example, taking climate change into account might tip the balance in support of the implementation of win-win measures which reduce energy use and improve indoor working conditions.

Many organisations are exposed to several direct and indirect consequences of climate change. Some adaptation measures can in principle address several risks simultaneously, but in other cases there may be conflicts between measures addressing different risks. Meanwhile, some risks may require the adoption of several adaptation measures. An adaptation strategy will therefore likely comprise a set of different adaptation measures.

The feasibility and suitability of different adaptation measures depend on the organisation and its context. It is therefore essential that comprehensive approaches are used to identify adaptation options, and that robust procedures are used to evaluate adaptation options according to an organisation's needs and circumstances.

DESIGNING AND APPRAISING ADAPTATION OPTIONS

There are, in principle, therefore many potential adaptation options available to an organisation to address risks and opportunities provided by climate change. In practice some will be infeasible, some will be too expensive, and some may be politically difficult to implement. This narrowing of adaptation from the range of all possible options to a small feasible subset is illustrated schematically in Figure 6.1.[3]

3 Chambwere, M., et al. (2014), The Economics of Adaptation. In: *Climate Change 2014: Impacts, Adaptation and Vulnerability. Contribution of Working Group II to the Fifth*

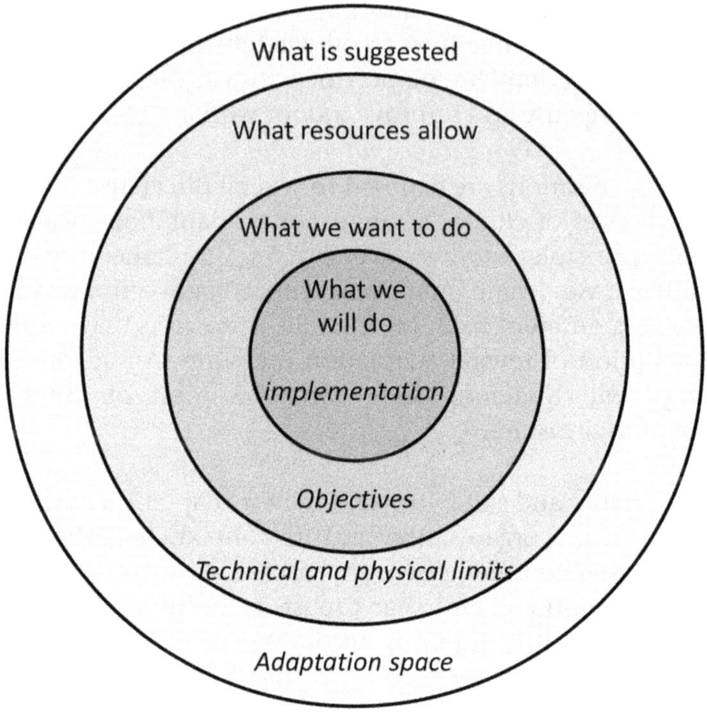

Figure 6.1 The narrowing of the adaptation space (from IPCC AR5 Working Group II Report, Chapter 17)

This section considers two main issues. The first is the set of criteria which could be used to evaluate different adaptation options, and the second the process used in option appraisal. As we shall see, some of these appraisal processes explicitly incorporate, through iteration, the design of adaptation options. There are typically two components to an adaptation decision – what to do and when to do it – and both therefore

Assessment Report of the Intergovernmental Panel on Climate Change. Field, C., et al., eds. Cambridge University Press.

need to be explicitly considered when looking at criteria and appraisal processes.

ADAPTATION OBJECTIVES AND EVALUATION CRITERIA

There are several potential objectives for adaptation. These include:

- Maintain risks at current levels;
- Maintain an economically optimal level of risk;
- Reduce exposure to risk compared with current levels;
- Improve resilience to future uncertainty;
- Exploit opportunities provided by climate change.

The difference between the first two objectives is subtle, but important. The first seeks to maintain the same standard of risk (for example likelihood of an event occurring or standard or service), whilst the second seeks to establish a new 'economically optimum' level of risk; this may be higher or lower than the current level of risk.

Given a specific objective, there are several criteria for the evaluation of individual potential adaptation options (Table 6.3). Some can be seen as values to be optimised (or minimised); others are better seen as constraints on adaptation.

Technical effectiveness represents the ability of the option actually to meet objectives: will it reduce risks, and can we be sure that it will work? There may be uncertainty about how

Table 6.3 Potential criteria for the evaluation of adaptation options

Criteria
Technical effectiveness
Technical feasibility
Economic effectiveness
Economic feasibility
'Political' feasibility
Availability of relevant information
Robustness to uncertainty
Flexibility
Residual impacts
Co-benefits
Consistency with other objectives

well novel approaches will work in practice, uncertainty over whether the option would cope with all potential futures, and uncertainty over whether the actions of agents inside and outside the organisation would limit the effectiveness of the option. For example, the effectiveness of a measure which involves reducing water demands would depend on the actions of individuals. Technical feasibility is different to effectiveness, as it describes whether an option is suitable for a specific circumstance and whether it can be implemented in practice. For example, a reliance on evacuation may not be feasible where warning times are short, and there are many technical constraints on structural adaptation options. Legal or regulatory factors may affect the technical feasibility of a potential option, as can the knowledge and expertise available to the organisation. Technical feasibility also covers timing and scheduling; an option may be technically feasible on one time-scale, but not on another. Many of the measures to enhance adaptive capacity outlined above seek to affect the technical feasibility of adaptation options.

Economic effectiveness describes the relationship between the economic benefits of an option and its costs, as represented for example by its benefit–cost ratio. Economic *feasibility*, in contrast, relates to the ability to raise the funds necessary to implement the option. An option may have a very high benefit–cost ratio, but have a large absolute cost and therefore be unaffordable.

'Political' feasibility (with a small p) describes the social acceptability of an adaptation option, and can pose substantial constraints. An adaptation option may require the delivery of services in a different way or to different standards which may not be acceptable to customers, suppliers or the general public. It may involve some construction work, which may be subject to local opposition.

The design of different options requires different amounts of information on the magnitude of potential climate change. At one extreme, some very 'technical' measures require specific quantitative information. For example, the design of a new cooling system for a building would require an estimate of future cooling requirements, and the design of a new flood protection wall would need information on the potential height of future floods. The amount of insurance to be bought would require an estimate of potential financial losses. At the other extreme, some measures do not need quantitative information on future climate change. For example, measures to increase water use efficiency do not need to be based on projected changes in river flows, and the design of an emergency plan to be activated following a warning does not need quantitative climate projections. Between these two extremes, different specific measures require different amounts of information at the design stage. For some adaptation options, the availability of credible and relevant information on which to base a design may be a significant constraint. Chapter 5 outlined the types of information that are typically available from climate scenarios; basically, the more detailed and specific the climate information needed to design the option, and the greater the importance of potential changes in short-duration or extreme events, the greater the constraints on adaptation posed by the availability of information.

The amount of climate change expected at a place is uncertain. An adaptation option designed to cope with one particular outcome may perform badly if the future climate is, in actuality, different to that expected. The robustness of an adaptation option to a range of different futures is therefore a potentially important criterion. Other things being equal, an option which works with a range of future outcomes will be better than an option that only works if climate change is within specific narrow bounds: it has an 'option value'. However, there is a balance between robustness and economic effectiveness. In a similar way, an option that requires only relatively basic climate change information is likely to be more robust than one which requires detailed and specific design information.

Climate change is not only uncertain, but our knowledge about how climate is changing evolves over time too. An adaptation option that is flexible and can be revised as more information becomes available is, again other things being equal, likely to be better than one which cannot be changed or upgraded once implemented. Examples include options which can be easily upgraded because capital stock is renewed rapidly, or options that can be designed with upgrade potential in mind.

Any adaptation option is unlikely to remove all the risks from climate change. Adaptation options targeted at extreme events are typically designed around some standard of protection. If events beyond this standard occur, they will generate losses or damage: these are known as 'residual impacts'. The magnitude of these residual impacts may vary with different adaptation options which have the same notional standard of protection. For example, a flood wall may provide complete protection up to a particular size flood, but no protection at all in larger floods. In contrast, measures to reduce losses during an event may reduce, even if not eliminate, losses in events larger than the design standard. The distinction between 'fail-safe' and 'safe-fail' designs is important here (the former designs not to fail, and the latter designs so that failure causes minimal problems).

Finally, organisations are rarely seeking to adapt to climate change alone. Important extra potential criteria are therefore the co-benefits that a given adaptation option generates in other areas (for example changing building design to reduce the impact of heatwaves and also reduce energy use), and consistency with other organisational objectives. It is possible that in some cases options to adapt to climate change may be inconsistent with other objectives.

APPRAISAL TOOLS

There are clearly therefore not only many potential adaptation options available to an organisation to address climate change risks and exploit opportunities, but also many criteria against which these options can be judged. It will be necessary to make explicit trade-offs between different criteria – against the background of uncertainty about how the climate will change at a place. A tiered approach to option appraisal can eliminate potential options from further consideration, based on the qualitative interpretation of the criteria in Table 6.3, leaving a subset for more detailed – typically economic – analysis.

The conventional approach to option appraisal, as used in corporate investment analysis, is to compare the stream of costs and benefits of each option under a specific set of assumptions about the future in a discounted cash flow (DCF) analysis. However, as has been emphasised many times, the future is uncertain so our projections of future costs and benefits are also uncertain. This makes our comparison of costs and benefits uncertain, and we may not be sure that benefits exceed costs. It also complicates the comparison of options because different options may be differently sensitive to uncertainty. The standard way of incorporating uncertainty into discounted cash flow analysis is via the discount rate, adding a risk premium. However, such a risk premium needs to be quantified, and this is extremely difficult in the context of uncertain future climate change.

Another approach to appraisal is based on concepts developed in the field of decision-making under uncertainty, using a payoff table. A payoff table summarises what would happen under a given decision if the future were to turn out in a certain way. Table 6.4a shows a hypothetical example, where we are concerned with an increase in flood risk. In this case, four adaptation options are being considered: the first three assume that measures are taken to eliminate impacts when the flood risk increases by 10, 20 or 30%, and the fourth assumes that nothing is done. In the example, we assume that adaptation costs 1 unit for every percentage increase that is prevented (so adapting to an increased risk of 10% costs 10 units). We also assume that an increase in flood risk costs 2 units for every percentage increase above the level of protection. With no adaptation, a 20% increase in risk therefore would cost 40 units, and with adaptation up to an increase in risk of 10%, a 20% increase in risk would cost 20 units in residual impacts. The payoff table shown in Table 6.4a shows the total cost of each adaptation option (cost of implementation plus residual impacts) under four different possible futures. The values are negative because they represent expenditure. For example, adaptation option 2 (adapt to a 20% increase in risk) would cost 40 units if the future were to see an increase in risk of 30% (a cost of 20 units for the adaptation option, plus 20 units for the residual impacts).

There are three potential rules for selecting an option from a payoff matrix. The first is known as the *maximax* rule, and states that the option which has the maximum potential payoff should be selected; this rule is appropriate for a risk-taker. In the example in Table 6.4a, the maximax rule would select Option 1; this has the smallest potential cost (-10) and hence the maximum potential payoff. The *maximin* rule states that the option which has the largest minimum potential payoff should be selected, and this is appropriate for a risk-averse decision-maker. In the example in Table 6.4a, the maximin rule would select Option 3: this option has the smallest maximum cost (-50 units, which would arise if risk increased by 40%). The third rule – the *minimax* rule – states that the option with the smallest maximum 'regret' should be selected. 'Regret' is the opportunity cost of making the wrong decision – or in other words not selecting the most appropriate option given what actually happens. This is presented in a regrets table (Table 6.4b). A regrets table is produced by first finding the maximum payoff (or, in our case, the minimum cost) across all options for a given future state, and then subtracting this from the payoff for each option. The option to be selected is that with the lowest maximum regret across all future states. For example, if the future state were to be an increase in risk of 10%, then the maximum payoff (or minimum cost) would be with adaptation option 1; option 2 would have a cost of -20, and option 3 a cost of -30. Selecting option 2 would therefore cost 10 units more than the 'best' option, and option 3 would cost 20 units more. If the future state were to be an increase in risk of 20%, then option 2 would be the best, and options 1, 3 and 4 would cost more. In the example in Table 6.4b, option 2 has the smallest maximum regret across the four possible future states. The minimax rule is appropriate for organisations that are risk-neutral, and is therefore the rule most frequently used in decision-making under uncertainty.

Table 6.4 is hypothetical and highly stylised, but illustrates the general principles of decision-making under uncertainty. The potential future states can be different specific quantitative scenarios (for example from different climate model simulations as described in Chapter 5), or can be specific prescribed changes in relevant drivers of change (such as an increase in risk of 10 or 20%). However, if such prescribed changes are used, then we need some way of defining what range of changes is plausible. In the case of Table 6.4, how realistic would an increase in risk of 40%, or indeed 10%, be? The assessment of the plausible range for future states in the payoff matrix could be based on climate scenarios, even if the individual states are not.

Another decision rule used in decision-making under uncertainty is to select the option with the *maximum expected payoff,* calculated by multiplying the payoff under each future state with the estimated probability of that future state occurring. However, as explained in

Table 6.4a Decision-making under uncertainty: a hypothetical payoff table

Actual future change	Option 1 (adapt to +10%)	Option 2 (adapt to +20%)	Option 3 (adapt to +30%)	Option 4 (do nothing)
10%	-10	-20	-30	-20
20%	-30	-20	-30	-40
30%	-50	-40	-30	-60
40%	-70	-60	-50	-80

Table 6.4b A hypothetical regrets table

Actual future change	Option 1 (adapt to +10%)	Option 2 (adapt to +20%)	Option 3 (adapt to +30%)	Option 4 (do nothing)
10%	0	10	20	10
20%	10	0	10	20
30%	20	10	0	30
40%	20	10	0	30

Chapter 5, it is very challenging to estimate robustly the likelihood of future climate change outcomes. We cannot assign likelihoods to individual scenarios, we cannot assign likelihoods to specific quantitative changes (only to ranges in change), and our estimated likelihoods are very dependent on the set of scenarios we use. It is therefore difficult to use the maximum expected payoff rule in the context of climate change.

In Table 6.4, each adaptation option implicitly represents perfect adaptation to one of the future states. This could be because each option is designed to cope with a specific scenario. However, this does not necessarily have to be the case.

There are three problems with the application of standard decision-making under uncertainty techniques and rules in the evaluation of options to adapt to climate change. The first is that the evaluation can be strongly constrained by the assumed future states that are considered – by the number of climate scenarios, or by the range of prescribed future states that are used. The second is that the standard approach assumes that a decision, once made, is fixed and unchangeable. In practice, many actions are revised as more information becomes available. Third, the standard approach is better suited to evaluating *what* should be done rather than *when* something should be done. These last two features are characteristic of adaptation to climate change.

Real options analysis (ROA) is an evaluation technique that is designed to cope with decisions that are based on learning, and where timing is important.[4] It was developed from approaches used to evaluate financial options. A 'real option' is simply defined as the option to make some decision on a management intervention in the future, and is distinct from a financial option which is an option to buy or sell something in the future. Financial options are often evaluated using theoretical models, but real options – which are by definition more 'real' and practical than financial options – are typically evaluated through the use of decision trees. A decision tree describes what would happen at a given time period if specific conditions were met; if there are just two options at each time period, then such a decision tree is frequently termed a binomial lattice. By examining the performance of

4 HM Treasury and Defra (2009), *Accounting for the effects of climate change. Supplementary Green Book Guidance*, June 2009. HM Treasury.

different options and their implementation (as defined by decision trees) over a range of possible futures, a real options analysis can be used to identify not only flexible adaptation options, but also options robust to uncertainty.

Table 6.5 provides a simple hypothetical example of the application of real options analysis, based on protection against flooding. In this example there are two potential options: a standard fixed wall costing 75 units, and an upgradeable wall costing 50 units initially – and providing less protection – but capable of being upgraded at a cost of 50 units to provide higher protection than the standard fixed wall if the amount of climate change passes some threshold. The impacts depend on the amount of climate change. If climate change is 'large', then impacts cost 100 units in time step 1 and 150 units in time step 2, and if climate change is 'small', there are no impacts in either time period. If the standard wall is built and climate change is 'large', then 100 units of damage are saved in time step 1 and 100 are saved in time step 2, but 50 units of damage still occur. The total benefits of the standard wall equal 75 units over the two time steps. If the upgradeable wall is built at a cost of 50 units, then 75 units of damage are saved in time period 1 if climate change is large, but 25 units of damage still occur. However, because the climate change is large the wall is upgraded by the end of time step 2 (at an additional cost of 50 units), saving 150 units of damage in time step 2. The total benefits of the upgradeable wall are therefore 100 units. If, on the other hand, climate change is small, then there are no benefits in either time period, but the standard wall incurs a cost of -75 units and the upgradeable wall a cost of only -50 units (because it is not upgraded). In this simple example, therefore, the upgradeable wall provides better value than the fixed wall, regardless of the actual amount of climate change that occurs, over the long term.

Table 6.5 A simple hypothetical real options analysis

		Time step 1			**Time step 2**			
Adaptation measure	Initial cost	Future world	Damages saved	Residual damages	Upgrade cost	Damages saved	Residual damages	**TOTAL**
Standard wall	-75	Large change	100	0	0	100	-50	**75**
	-75	Small change	0	0	0	0	0	**-75**
Upgradeable wall	-50	Large change	75	-25	50	150	0	**100**
	-50	Small change	0	0	0	0	0	**-50**

The example in Table 6.5 is of course again highly stylised. In practice, there will be more potential adaptation options, more ways of scheduling the implementation of each option, more potential future climates, and the analysis will be undertaken over more time periods. Real options analysis can therefore be challenging to implement, as it will probably involve the calculation of outcomes under many combinations of circumstances. It also needs climate scenarios defined for successive time periods.

The techniques outlined so far are designed to evaluate the economic effectiveness of different adaptation options. There are, as discussed in the previous section, many other potentially important criteria for evaluation. These could be used to screen out 'infeasible' options prior to an economic analysis, but this can rather downplay the importance of non-economic criteria in the evaluation of adaptation options. One way of using a large number of criteria to evaluate options is to undertake a *multicriteria analysis* (MCA). This involves defining the set of criteria against which options are to be evaluated, and producing ways of scoring performance against these criteria. The scoring can be quantitative or qualitative (for example based on expert judgement), but it is important to ensure that the scoring rules are consistent across the criteria. Some applications of MCA produce composite scores across all criteria to evaluate options, often weighting the different criteria differently. It is essential in the application of MCA that scoring and weighting rules are transparent, and it is also important that the effects of uncertainty in the scoring and weighting on the overall ranking of options are assessed.

Another potential multi-criteria approach to the evaluation of adaptation options is to conduct a 'stress-test' in order to see how the adaptation strategy works under 'extreme' conditions. The stress test could consider not only economic aspects of the adaptation strategy, but also practical issues around implementation and any potential co-benefits or unanticipated adverse consequences. Stress testing is, however, only as credible as the 'extreme' scenarios that are considered.

So far, this section has outlined a few methods for evaluating the performance of a set of well-specified adaptation options. It is also possible to use the evaluation process to refine and develop adaptation options in an iterative sense. For example, it might be possible to adjust an adaptation option designed around one specific climate scenario so that it also performs well against others. A formalised way of doing this is through the technique of *robust decision-making*.[5] Robust decision-making starts with a potential adaptation option or strategy, and then evaluates it (against defined criteria) against a large number of climate scenarios. The performance of the adaptation option is then compared with key dimensions of these climate scenarios. These could be, for example, the change in summer temperature or change in amount of winter rainfall. If the performance of the adaptation option depends very much on some specific aspects of climate change, then the adaptation option is adjusted to reduce sensitivity, and its performance again evaluated with a large number of scenarios. For example, if it is found that performance depends on change in winter precipitation, the adaptation option is revised to better cope with uncertainty in future winter precipitation. The process can be repeated several times, and in principle homes in on an adaptation option or strategy that is robust to uncertainty (or is as robust as feasible). The process is intensive and

5 Weaver, C.P., et al. (2013), Improving the contribution of climate model information to decision-making: the value and demands of robust decision frameworks. *WIRES: Climate Change* 4, 39-60.

requires a large number of climate scenarios, a capability to simulate large numbers of possible outcomes, and an ability to learn and revise adaptation options.

MONITORING AND EVALUATION

Monitoring and evaluation are key components of any risk management framework. The aim is to review options as implemented and their performance, and adjust if necessary. Monitoring typically involves measuring the performance of an intervention and comparing with pre-defined targets or standards. Climate change poses specific challenges to monitoring and evaluation.

Adapting to climate change is primarily concerned with anticipating future risks and opportunities. These may not be realised for years (or even decades), and it is therefore difficult to monitor performance in reducing risks and increasing opportunities. An exception is where the adaptation strategy is seeking to reduce exposure to current vulnerability: there it is possible to learn from experience. More generally, there are three types of monitoring in the face of climate change.

The first is to monitor *projections* of potential future climate changes, in order to decide whether to adjust the adaptation strategy or individual options. This is an essential task because of the significance of uncertainty in the projected consequences of climate change. Over time, new projections of plausible climate changes will be developed as understanding of the climate system improves, as new techniques become available, and as accumulated experience allows the comparison of projections with observations. New information will not necessarily reduce the range of uncertainty in the projected consequences of climate change at a place. It is quite possible that greater understanding of the climate system and how it impacts on the physical environment would lead to increased awareness of the potential for a wider range of consequences. In the UK, for example, the UKCP09 climate projections span a greater range of potential climate changes than the previous UKCIP02 scenarios, primarily because they are based on a larger number of climate models. It should not be assumed that any climate projections used to design and evaluate adaptation options will remain appropriate through the life of the adaptation option.

The second is to monitor the *implementation* of an adaptation strategy or option, rather than its performance in achieving adaptation targets. This monitoring would consider whether the implementation is proceeding on schedule and to cost, and seek to identify any 'surprises' or unanticipated barriers or knock-on consequences.

The third is to monitor *changing barriers* to adaptation. These can include changes in the regulatory environment which make it more or less difficult to adapt to climate change, and changes in the availability or feasibility of specific adaptation options. It also includes monitoring the adaptive capacity of the organisation, particularly the access of the organisation to knowledge and skills.

All these types of monitoring and evaluation are vital components of an adaptation strategy. Given that the future is uncertain, adaptation is about learning and revising strategies and actions. Monitoring and evaluation could be undertaken on an *ad hoc* basis (for example following an event), but is more appropriately done at fixed intervals as a specific and explicit part of the adaptation strategy. These intervals may be determined by relevant time horizons in the regulatory process (as part of regulatory periodic review or licence renewal), or by dates at which new information will become available. For example, reviews may be timed to coincide with the availability of new sets of climate scenarios, or the publication of future IPCC or national assessment reports.

CONCLUSION

This chapter has introduced the key issues involved in managing the risks posed, and opportunities provided, by climate change. The details of how an organisation will and could adapt to climate change depend on the context within which that organisation operates, but there are several general points.

Adaptation to climate change can be seen as a specific task, or as one dimension of a more general review or investment plan. There are different approaches to adaptation, all with advantages and disadvantages. Improving resilience to current climatic vulnerability, for example, is one approach but does not guarantee improved resilience to future climate change. 'Predict and plan' is widely used in many investment decisions but a key feature of climate change is its uncertainty. Adaptive planning deals with uncertainty, but is not necessarily very efficient for long-lived investments or where options are difficult to upgrade. There can be relationships – co-benefits or tensions – between measures to adapt to climate change and measures to reduce emissions.

The uncertainty in how climate change might impact upon an organisation is central to the development and appraisal of adaptation strategies and options. Learning and flexibility are important, and robustness to uncertainty is a key evaluation criterion. Conventional business planning and investment tools are not well suited to climate change adaptation as they typically assume that the future can be predicted and that decisions, once made, are fixed. Tools such as real options analysis are designed to cope with uncertainty and flexibility, and iterative procedures such as robust decision-making can help in the development of adaptation strategies and options that are robust to uncertainty.

However, these tools are more demanding – in terms of skills, resources and knowledge – than conventional tools, and the concept of undertaking a tiered assessment of adaptation options is important. The level of detail required when defining and evaluating adaptation depends on the characteristics and magnitude of the potential climate change risk, and the level of detail that is feasible depends on the availability of plausible and credible information on climate change and its consequences. Nevertheless, it is important to take a systematic approach in a tiered assessment to ensure that all potential options and criteria are considered.

Finally, it is important to recognise the many barriers to adaptation. Some of these – particularly relating to the external environment in which the organisation operates – may be difficult to overcome, but many are a feature of the organisation and its procedures. Reducing these barriers, through increasing adaptive capacity, is an extremely important specific adaptation option for an organisation.

7 An Overview

There is now very strong evidence that the earth's climate is changing, and that this is largely due to human activity – mostly the use of fossil fuels. Temperatures and sea levels have risen, the frequency of heavy rainfall events has increased in many areas, and natural systems in many parts of the world are changing due to the changes in climate. We can no longer assume that the past is a good guide when we are assessing environmental risks. Whilst we cannot blame all individual extreme events on climate change, we can suspect that the dice are already increasingly loaded.

Climate change will accelerate, and we can expect to see more impacts on both natural and human systems in the future. Reducing climate change requires substantial and sustained reductions in emissions, and even if that were to happen we will not eliminate all impacts.

Climate change therefore has a very strong potential to alter business risks, and also to provide new opportunities. It affects operational, supply chain, market, regulatory and reputational risks, although how it does so depends on the characteristics of an organisation. Risk is determined not only by the amount of climate change, but also by exposure and vulnerability. Organisations need to seriously consider whether climate change should be added to their risk register.

However, organisations can seek to adapt to climate change in order to reduce risks and exploit opportunities. They can implement specific measures to deal just with climate change, or they can incorporate climate change into their usual strategic and investment planning processes alongside other changes and drivers. Adapting to climate change is challenging, for a number of reasons.

Firstly, we do not know exactly how climate will change at a place. We can make projections of plausible changes, but these can be quantitatively uncertain for several reasons. We cannot predict exactly how future emissions will change, because they depend on future energy use and, to a certain extent, on policy choices. We understand well the basic physics behind climate change, but the actual consequences at a place are very difficult to estimate because our ability to understand and represent the details of how climate works is constrained. Finally, the climate system is subject to internal variability, and over the next few years this may disguise the effects at a place of underlying climate change. These sources of uncertainty are important at different time-scales, but combine to mean that we cannot forecast future climate; we can only make physically plausible projections. We may also find it challenging to estimate relevant indicators of the impact of a given change in weather and climate. However, through the use of multiple scenarios and models, we can get some idea of the potential range in impacts at a place.

Second, the impacts of climate change may be overwhelmed, dampened or exaggerated by other changes that are taking place, and changes to the market and regulatory environment may affect how an organisation can respond to risk and exploit opportunities.

Third, although climate change is already affecting some risks, the impacts of climate change are likely to be most apparent and identifiable in the coming years and decades. These future impacts may be beyond the time-scales normally considered by an organisation – but may actually affect the outcome of decisions made now.

Fourth, conventional business investment appraisal approaches are not well suited to climate change, partly because of the uncertainty in how climate will change in the future, and partly because our assessment of what may happen will change over time as we gain more information. We typically cannot therefore seek to 'predict and plan', and have to implement more flexible and adaptable approaches. These require the use of different risk assessment and option appraisal techniques, such as scenario planning and techniques for decision-making under uncertainty, as well as more novel approaches such as real options analysis and robust decision-making.

Finally, adaptation requires information on an organisation's exposure and vulnerability to climate change, and on how relevant aspects of weather and climate may change in the future. This information may not be available, either because an organisation does not understand enough about its exposure and vulnerability (especially if it has not recently experienced challenging events), or because plausible changes in the precise aspects of weather and climate that are relevant are difficult to estimate or have wide uncertainty bands. In general, the more extreme the relevant weather variable, the greater the uncertainty in plausible changes in that variable. This lack of information can influence the type of adaptation that can take place.

The response of an organisation to the risks and opportunities of climate change is very much determined by the nature of that organisation and the challenges it faces. However, it is possible to identify a number of general principles an organisation should consider when addressing climate change:

- Develop the organisation's capacity to adapt, by developing skills and information and seeking to influence external barriers.
- Understand current exposure and vulnerability to weather and climate variability.
- Improve the organisation's resilience to weather and climatic variability.
- Seek to implement actions that are robust to uncertainty and flexible enough to be revised or updated, whilst remaining economically effective.
- Avoid locking in to long-term decisions that cannot be revised or updated.
- Avoid taking decisions which increase future exposure and vulnerability to weather events and climate change.
- Continue to learn from experience and the evolving understanding of climate change and its consequences.
- Exploit the opportunities provided by setting multiple objectives to achieve multiple co-benefits.

Climate change is a real and emerging risk to organisations, which challenges the way they identify, evaluate and manage risks. However – as this book has outlined – these risks can be managed and opportunities that present themselves can be exploited.

Index

www.ingramcontent.com/pod-product-compliance
Lightning Source LLC
LaVergne TN
LVHW011127300926
849160LV00008B/550
* 9 7 8 1 4 0 9 4 5 3 5 2 9 *